PROPERTY OF THE LIBRARY
ASSUMPTION COLLEGE
BRANTFORD, ONTARIO

*THE MEANING OF RELATIVITY*

THE STAFFORD LITTLE LECTURES
OF PRINCETON UNIVERSITY
MAY 1921

PROPERTY OF THE LIBRARY
ASSUMPTION COLLEGE
BRANTFORD, ONTARIO
JUN - - 1994

# THE MEANING OF RELATIVITY

*Fifth edition,*
*including the*

*RELATIVISTIC THEORY OF THE NON-SYMMETRIC FIELD*

*By* ALBERT EINSTEIN

INSTITUTE FOR ADVANCED STUDY

PRINCETON UNIVERSITY PRESS
PRINCETON, NEW JERSEY

The first edition of this book, published in 1922 by Methuen and Company in Great Britain and by Princeton University Press in the United States, consisted of the text of Mr. Einstein's Stafford Little Lectures, delivered in May 1921 at Princeton University. For the second edition, Mr. Einstein added an appendix discussing certain advances in the theory of relativity since 1921. For the third edition, Mr. Einstein added an appendix (Appendix II) on his Generalized Theory of Gravitation. This was completely revised for the fifth edition.

COPYRIGHT © 1922, 1945, 1950, 1953
BY PRINCETON UNIVERSITY PRESS

COPYRIGHT © 1956, BY THE ESTATE
OF ALBERT EINSTEIN

COPYRIGHT RENEWED 1984 BY THE HEBREW
UNIVERSITY OF JERUSALEM

L.C. Card: 56-1198

ISBN 0-691-08007-0 (hardcover edn.)

ISBN 0-691-02352-2 (paperback edn.)

PRINTED IN THE UNITED STATES OF AMERICA
BY PRINCETON UNIVERSITY PRESS, PRINCETON, N.J.

Princeton Science Library edition, 1988

19 18 17 16 15 14 13 12 11

19 18 17 16 15 (pbk.)

THE TEXT OF THE FIRST EDITION WAS TRANSLATED BY EDWIN PLIMPTON ADAMS, APPENDIX I BY ERNST G. STRAUS, APPENDIX II BY SONJA BARGMANN

This book is sold subject to the condition that it shall not, by way of trade, be lent, resold, hired out, or otherwise disposed of without the publisher's consent, in any form of binding or cover other than that in which it is published.

# CONTENTS

# A NOTE ON THE FIFTH EDITION

For the present edition I have completely revised the "Generalization of Gravitation Theory" under the title "Relativistic Theory of the Non-symmetric Field." For I have succeeded—in part in collaboration with my assistant B. Kaufman—in simplifying the derivations as well as the form of the field equations. The whole theory becomes thereby more transparent, without changing its content.

A. E.—December 1954

# SPACE AND TIME IN PRE-RELATIVITY PHYSICS

THE theory of relativity is intimately connected with the theory of space and time. I shall therefore begin with a brief investigation of the origin of our ideas of space and time, although in doing so I know that I introduce a controversial subject. The object of all science, whether natural science or psychology, is to co-ordinate our experiences and to bring them into a logical system. How are our customary ideas of space and time related to the character of our experiences?

The experiences of an individual appear to us arranged in a series of events; in this series the single events which we remember appear to be ordered according to the criterion of "earlier" and "later," which cannot be analysed further. There exists, therefore, for the individual, an I-time, or subjective time. This in itself is not measurable. I can, indeed, associate numbers with the events, in such a way that a greater number is associated with the later event than with an earlier one; but the nature of this association may be quite arbitrary. This association I can define by means of a clock by comparing the order of events furnished by the clock with the order of the given series of events. We understand by a clock something which provides a series of events which can be counted, and which has other properties of which we shall speak later.

By the aid of language different individuals can, to a certain extent, compare their experiences. Then it turns out that certain sense perceptions of different individuals correspond to each other, while for other sense perceptions no such correspondence can be established. We are accustomed to regard as real those sense perceptions which are common to different individuals, and which therefore are, in a measure, impersonal. The natural sciences, and in particular, the most fundamental of them, physics, deal with such sense perceptions. The conception of physical bodies, in particular of rigid bodies, is a relatively constant complex of such sense perceptions. A clock is also a body, or a system, in the same sense, with the additional property that the series of events which it counts is formed of elements all of which can be regarded as equal.

The only justification for our concepts and system of concepts is that they serve to represent the complex of our experiences; beyond this they have no legitimacy. I am convinced that the philosophers have had a harmful effect upon the progress of scientific thinking in removing certain fundamental concepts from the domain of empiricism, where they are under our control, to the intangible heights of the *a priori*. For even if it should appear that the universe of ideas cannot be deduced from experience by logical means, but is, in a sense, a creation of the human mind, without which no science is possible, nevertheless this universe of ideas is just as little independent of the nature of our experiences as clothes are of the form of the human body. This is particularly true of our concepts of time and space, which physicists have been obliged by the facts to bring down from the Olympus of the *a priori* in order to adjust them and put them in a serviceable condition.

We now come to our concepts and judgments of space. It is essential here also to pay strict attention to the relation of experience to our concepts. It seems to me that Poincaré clearly recognized the truth in the account he gave in his book, "La Science et l'Hypothese." Among all the changes which we can perceive in a rigid body those which can be cancelled by a voluntary motion of our body are marked by their simplicity; Poincaré calls these, changes in position. By means of simple changes in position we can bring two bodies into contact. The theorems of congruence, fundamental in geometry, have to do with the laws that govern such changes in position. For the concept of space the following seems essential. We can form new bodies by bringing bodies $B$, $C$, . . . up to body $A$; we say that we *continue* body $A$. We can continue body $A$ in such a way that it comes into contact with any other body, $X$. The ensemble of all continuations of body $A$ we can designate as the "space of the body $A$." Then it is true that all bodies are in the "space of the (arbitrarily chosen) body $A$." In this sense we cannot speak of space in the abstract, but only of the "space belonging to a body $A$." The earth's crust plays such a dominant rôle in our daily life in judging the relative positions of bodies that it has led to an abstract conception of space which certainly cannot be defended. In order to free ourselves from this fatal error we shall speak only of "bodies of reference," or "space of reference." It was only through the theory of general relativity that refinement of these concepts became necessary, as we shall see later.

I shall not go into detail concerning those properties of the space of reference which lead to our conceiving points as elements of space, and space as a continuum. Nor shall I attempt to analyse further the properties of space which justify the conception of continuous series of points, or lines. If these concepts are assumed, together with their relation to the solid bodies of experience, then it is easy to say what we mean by the three-dimensionality of space; to each point three numbers, $x_1$, $x_2$, $x_3$ (co-ordinates), may be associated, in such a way that this association is uniquely reciprocal, and that $x_1$, $x_2$, and $x_3$ vary continuously when the point describes a continuous series of points (a line).

It is assumed in pre-relativity physics that the laws of the configuration of ideal rigid bodies are consistent with Euclidean geometry. What this means may be expressed as follows: Two points marked on a rigid body form an *interval.* Such an interval can be oriented at rest, relatively to our space of reference, in a multiplicity of ways. If, now, the points of this space can be referred to co-ordinates $x_1$, $x_2$, $x_3$, in such a way that the differences of the co-ordinates, $\Delta x_1$, $\Delta x_2$, $\Delta x_3$, of the two ends of the interval furnish the same sum of squares,

$$s^2 = \Delta x_1{}^2 + \Delta x_2{}^2 + \Delta x_3{}^2 \quad (1)$$

for every orientation of the interval, then the space of reference is called Euclidean, and the co-ordinates Cartesian.* It is sufficient, indeed, to make this assumption in the limit for an infinitely small interval. Involved in this assumption there are some which are rather less special, to which we must call attention on account of their fundamental significance. In the first place, it is assumed that one can move an ideal rigid body in an arbitrary manner. In the second place, it is assumed that the behaviour of ideal rigid bodies towards orientation is independent of the material of the bodies and their changes of position, in the sense that if two intervals can

* This relation must hold for an arbitrary choice of the origin and of the direction (ratios $\Delta x_1 : \Delta x_2 : \Delta x_3$) of the interval.

once be brought into coincidence, they can always and everywhere be brought into coincidence. Both of these assumptions, which are of fundamental importance for geometry and especially for physical measurements, naturally arise from experience; in the theory of general relativity their validity needs to be assumed only for bodies and spaces of reference which are infinitely small compared to astronomical dimensions.

The quantity $s$ we call the length of the interval. In order that this may be uniquely determined it is necessary to fix arbitrarily the length of a definite interval; for example, we can put it equal to 1 (unit of length). Then the lengths of all other intervals may be determined. If we make the $x_\nu$ linearly dependent upon a parameter $\lambda$,

$$x_\nu = a_\nu + \lambda b_\nu,$$

we obtain a line which has all the properties of the straight lines of the Euclidean geometry. In particular, it easily follows that by laying off $n$ times the interval $s$ upon a straight line, an interval of length $n \cdot s$ is obtained. A length, therefore, means the result of a measurement carried out along a straight line by means of a unit measuring rod. It has a significance which is as independent of the system of co-ordinates as that of a straight line, as will appear in the sequel.

We come now to a train of thought which plays an analogous rôle in the theories of special and general relativity. We ask the question: besides the Cartesian co-ordinates which we have used are there other equivalent co-ordinates? An interval has a physical meaning which is independent of the choice of co-ordinates; and so has the spherical surface which we obtain as the locus of the end points of all equal intervals that we lay off from an arbitrary point of our space of reference. If $x_\nu$ as well as $x'_\nu$ ($\nu$ from 1 to 3) are Cartesian co-ordinates of our space of reference, then the spherical surface will be expressed in our two systems of co-ordinates by the equations

$$\sum \Delta x_\nu{}^2 = \text{const.} \tag{2}$$

$$\sum \Delta x'_\nu{}^2 = \text{const.} \tag{2a}$$

How must the $x'_\nu$ be expressed in terms of the $x_\nu$ in order that equations (2) and (2a) may be equivalent to each other? Regarding the $x'_\nu$ expressed as functions of the $x_\nu$, we can write, by Taylor's theorem, for small values of the $\Delta x_\nu$,

$$\Delta x'_\nu = \sum_\alpha \frac{\partial x'_\nu}{\partial x_\alpha} \Delta x_\alpha + \frac{1}{2} \sum_{\alpha\beta} \frac{\partial^2 x'_\nu}{\partial x_\alpha \partial x_\beta} \Delta x_\alpha \Delta x_\beta \ldots$$

If we substitute (2a) in this equation and compare with (1), we see that the $x'_\nu$ must be linear functions of the $x_\nu$. If we therefore put

$$x'_\nu = \alpha_\nu + \sum_\alpha b_{\nu\alpha} x_\alpha \tag{3}$$

or

$$\Delta x'_\nu = \sum_\alpha b_{\nu\alpha} \Delta x_\alpha \tag{3a}$$

then the equivalence of equations (2) and (2a) is expressed in the form

$$\sum \Delta x'_\nu{}^2 = \lambda \sum \Delta x_\nu{}^2 \quad (\lambda \text{ independent of } \Delta x_\nu) \tag{2b}$$

It therefore follows that $\lambda$ must be a constant. If we put $\lambda = 1$, (2b) and (3a) furnish the conditions

$$\sum_\nu b_{\nu\alpha} b_{\nu\beta} = \delta_{\alpha\beta} \tag{4}$$

in which $\delta_{\alpha\beta} = 1$, or $\delta_{\alpha\beta} = 0$, according as $\alpha = \beta$ or $\alpha \neq \beta$. The conditions (4) are called the conditions of orthogonality, and the transformations (3), (4), linear orthogonal transformations. If we stipulate that $s^2 = \sum \Delta x_\nu{}^2$ shall be equal to the square of the length in every system of co-ordinates, and if we always measure with the same unit scale, then $\lambda$ must be equal to 1. Therefore the linear orthogonal transformations are the only ones by means of which we can pass from one Cartesian system of co-ordinates in our space of reference to another. We see that in applying such transformations the equations of a straight line become equations of a straight line. Reversing equations (3a) by multiplying both sides by $b_{\nu\beta}$ and summing for all the $\nu$'s, we obtain

$$\sum b_{\nu\beta}\Delta x'_\nu = \sum_{\nu\alpha} b_{\nu\alpha}b_{\nu\beta}\Delta x_\alpha = \sum_{\alpha} \delta_{\alpha\beta}\Delta x_\alpha = \Delta x_\beta \tag{5}$$

The same coefficients, $b$, also determine the inverse substitution of $\Delta x_\nu$. Geometrically, $b_{\nu\alpha}$ is the cosine of the angle between the $x'_\nu$ axis and the $x_\alpha$ axis.

To sum up, we can say that in the Euclidean geometry there are (in a given space of reference) preferred systems of co-ordinates, the Cartesian systems, which transform into each other by linear orthogonal transformations. The distance $s$ between two points of our space of reference, measured by a measuring rod, is expressed in such co-ordinates in a particularly simple manner. The whole of geometry may be founded upon this conception of distance. In the present treatment, geometry is related to actual things (rigid bodies), and its theorems are statements concerning the behaviour of these things, which may prove to be true or false.

One is ordinarily accustomed to study geometry divorced from any relation between its concepts and experience. There are advantages in isolating that which is purely logical and independent of what is, in principle, incomplete empiricism. This is satisfactory to the pure mathematician. He is satisfied if he can deduce his theorems from axioms correctly, that is, without errors of logic. The questions as to whether Euclidean geometry is true or not does not concern him. But for our purpose it is necessary to associate the fundamental concepts of geometry with natural objects; without such an association geometry is worthless for the physicist. The physicist is concerned with the question as to whether the theorems of geometry are true or not. That Euclidean geometry, from this point of view, affirms something more than the mere deductions derived logically from definitions may be seen from the following simple consideration.

Between $n$ points of space there are $\frac{n(n-1)}{2}$ distances, $s_{\mu\nu}$; between these and the $3n$ co-ordinates we have the relations

$$s_{\mu\nu}^2 = (x_{1(\mu)} - x_{1(\nu)})^2 + (x_{2(\mu)} - x_{2(\nu)})^2 + \dots$$

From these $\frac{n(n-1)}{2}$ equations the $3n$ co-ordinates may be eliminated, and from this elimination at least $\frac{n(n-1)}{2} - 3n$ equations in the $s_{\mu\nu}$ will result.* Since the $s_{\mu\nu}$ are measurable quantities, and by definition are independent of each other, these relations between the $s_{\mu\nu}$ are not necessary *a priori*.

* In reality there are $\frac{n(n-1)}{2} - 3n + 6$ equations.

From the foregoing it is evident that the equations of transformation (3), (4) have a fundamental significance in Euclidean geometry, in that they govern the transformation from one Cartesian system of co-ordinates to another. The Cartesian systems of co-ordinates are characterized by the property that in them the measurable distance between two points, $s$, is expressed by the equation

$$s^2 = \sum \Delta x_\nu^2.$$

If $K_{(x_\nu)}$ and $K'_{(x_\nu)}$ are two Cartesian systems of co-ordinates, then

$$\sum \Delta x_\nu^2 = \sum \Delta x'^2_\nu.$$

The right-hand side is identically equal to the left-hand side on account of the equations of the linear orthogonal transformation, and the right-hand side differs from the left-hand side only in that the $x_\nu$ are replaced by the $x'_\nu$. This is expressed by the statement that $\sum \Delta x_\nu^2$ is an invariant with respect to linear orthogonal transformations. It is evident that in the Euclidean geometry only such, and all such, quantities have an objective significance, independent of the particular choice of the Cartesian co-ordinates, as can be expressed by an invariant with respect to linear orthogonal transformations. This is the reason that the theory of invariants, which has to do with the laws that govern the form of invariants, is so important for analytical geometry.

As a second example of a geometrical invariant, consider a volume. This is expressed by

$$V = \int\int\int dx_1 dx_2 dx_3.$$

By means of Jacobi's theorem we may write

$$\int\int\int dx'_1 dx'_2 dx'_3 = \int\int\int \frac{\partial(x'_1, x'_2, x'_3)}{\partial(x_1, x_2, x_3)} dx_1 dx_2 dx_3$$

where the integrand in the last integral is the functional determinant of the $x'_\nu$ with respect to the $x_\nu$, and this by (3) is equal to the determinant $|b_{\mu\nu}|$ of the coefficients of substitution, $b_{\nu\alpha}$. If we form the determinant of the $\delta_{\mu\alpha}$ from equation (4), we obtain, by means of the theorem of multiplication of determinants,

$$(6) \qquad 1 = \left|\delta_{\alpha\beta}\right| = \left|\sum_\nu b_{\nu\alpha} b_{\nu\beta}\right| = \left|b_{\mu\nu}\right|^2; \left|b_{\mu\nu}\right| = \pm 1$$

If we limit ourselves to those transformations which have the determinant $+1$* (and only these arise from continuous variations of the systems of co-ordinates) then $V$ is an invariant.

Invariants, however, are not the only forms by means of which we can give expression to the independence of the particular choice of the Cartesian co-ordinates. Vectors and tensors are other forms of expression. Let us express the fact that the point with the current co-ordinates $x_\nu$ lies upon a straight line. We have

$$x_\nu - A_\nu = \lambda B_\nu \text{ ($\nu$ from 1 to 3).}$$

Without limiting the generality we can put

$$\sum B_\nu^2 = 1.$$

* There are thus two kinds of Cartesian systems which are designated as "right-handed" and "left-handed" systems. The difference between these is familiar to every physicist and engineer. It is interesting to note that these two kinds of systems cannot be defined geometrically, but only the contrast between them.

If we multiply the equations by $b_{\beta\nu}$ (compare (3a) and (5)) and sum for all the $\nu$'s, we get

$$x'_\beta - A'_\beta = \lambda B'_\beta$$

where we have written

$$B'_\beta = \sum_\nu b_{\beta\nu} B_\nu; \; A'_\beta = \sum_\nu b_{\beta\nu} A_\nu.$$

These are the equations of straight lines with respect to a second Cartesian system of co-ordinates $K'$. They have the same form as the equations with respect to the original system of co-ordinates. It is therefore evident that straight lines have a significance which is independent of the system of co-ordinates. Formally, this depends upon the fact that the quantities $(x_\nu - A_\nu) - \lambda B_\nu$ are transformed as the components of an interval, $\Delta x_\nu$. The ensemble of three quantities, defined for every system of Cartesian co-ordinates, and which transform as the components of an interval, is called a vector. If the three components of a vector vanish for one system of Cartesian co-ordinates, they vanish for all systems, because the equations of transformation are homogeneous. We can thus get the meaning of the concept of a vector without referring to a geometrical representation. This behaviour of the equations of a straight line can be expressed by saying that the equation of a straight line is co-variant with respect to linear orthogonal transformations.

We shall now show briefly that there are geometrical entities which lead to the concept of tensors. Let $P_0$ be the centre of a surface of the second degree, $P$ any point on the surface, and $\xi_\nu$ the projections of the interval $P_0P$ upon the co-ordinate axes. Then the equation of the surface is

$$\sum a_{\mu\nu}\xi_\mu\xi_\nu = 1.$$

In this, and in analogous cases, we shall omit the sign of summation, and understand that the summation is to be carried out for those indices that appear twice. We thus write the equation of the surface

$$a_{\mu\nu}\xi_\mu\xi_\nu = 1.$$

The quantities $a_{\mu\nu}$ determine the surface completely, for a given position of the centre, with respect to the chosen system of Cartesian co-ordinates. From the known law of transformation for the $\xi_\nu$ (3a) for linear orthogonal transformations, we easily find the law of transformation for the $a_{\mu\nu}$*:

$$a'_{\sigma\tau} = b_{\sigma\mu}b_{\tau\nu}a_{\mu\nu}.$$

This transformation is homogeneous and of the first degree in the $a_{\mu\nu}$. On account of this transformation, the $a_{\mu\nu}$ are called components of a tensor of the second rank (the latter on account of the double index). If all the components, $a_{\mu\nu}$, of a tensor with respect to any system of Cartesian co-ordinates vanish, they vanish with respect to every other Cartesian system. The form and the position of the surface of the second degree is described by this tensor ($a$).

Tensors of higher rank (number of indices) may be defined analytically. It is possible and advantageous to regard vectors as tensors of rank 1, and invariants (scalars) as tensors of rank 0. In this respect, the problem of the theory of invariants may be so formulated: according to

* The equation $a'_{\sigma\tau}\xi'_\sigma\xi'_\tau = 1$ may, by (5), be replaced by $a'_{\sigma\tau}b_{\mu\sigma}b_{\nu\tau}\xi_\sigma\xi_\tau = 1$, from which the result stated immediately follows.

what laws may new tensors be formed from given tensors? We shall consider these laws now, in order to be able to apply them later. We shall deal first only with the properties of tensors with respect to the transformation from one Cartesian system to another in the same space of reference, by means of linear orthogonal transformations. As the laws are wholly independent of the number of dimensions, we shall leave this number, $n$, indefinite at first.

*Definition.* If an object is defined with respect to every system of Cartesian co-ordinates in a space of reference of $n$ dimensions by the $n^\alpha$ numbers $A_{\mu\nu\rho} \ldots$ ($\alpha$ = number of indices), then these numbers are the components of a tensor of rank $\alpha$ if the transformation law is

$$A'_{\mu'\nu'\rho'} \ldots = b_{\mu'\mu} b_{\nu'\nu} b_{\rho'\rho} \ldots A_{\mu\nu\rho} \ldots \tag{7}$$

*Remark.* From this definition it follows that

$$A_{\mu\nu\rho} \ldots B_\mu C_\nu D_\rho \ldots \tag{8}$$

is an invariant, provided that $(B)$, $(C)$, $(D)$ . . . are vectors. Conversely, the tensor character of $(A)$ may be inferred, if it is known that the expression (8) leads to an invariant for an arbitrary choice of the vectors $(B)$, $(C)$, etc.

*Addition and Subtraction.* By addition and subtraction of the corresponding components of tensors of the same rank, a tensor of equal rank results:

$$A_{\mu\nu\rho} \ldots \pm B_{\mu\nu\rho} \ldots = C_{\mu\nu\rho} \ldots \tag{9}$$

The proof follows from the definition of a tensor given above.

*Multiplication.* From a tensor of rank $\alpha$ and a tensor of rank $\beta$ we may obtain a tensor of rank $\alpha + \beta$ by multiplying all the components of the first tensor by all the components of the second tensor:

$$T_{\mu\nu\rho\ldots\alpha\beta\gamma\ldots} = A_{\mu\nu\rho\ldots} B_{\alpha\beta\gamma\ldots} \tag{10}$$

*Contraction.* A tensor of rank $\alpha - 2$ may be obtained from one of rank $\alpha$ by putting two definite indices equal to each other and then summing for this single index:

$$T_{\rho\ldots} = A_{\mu\mu\rho\ldots} \left( = \sum_{\mu} A_{\mu\mu\rho\ldots} \right) \tag{11}$$

The proof is

$$\begin{aligned} A'_{\mu\mu\rho\ldots} = b_{\mu\alpha} b_{\mu\beta} b_{\rho\gamma} \ldots A_{\alpha\beta\gamma\ldots} &= \delta_{\alpha\beta} b_{\rho\gamma} \ldots A_{\alpha\beta\gamma\ldots} \\ &= b_{\rho\gamma} \ldots A_{\alpha\alpha\gamma\ldots} \end{aligned}$$

In addition to these elementary rules of operation there is also the formation of tensors by differentiation ("Erweiterung"):

$$T_{\mu\nu\rho\ldots\alpha} = \frac{\partial A_{\mu\nu\rho\ldots}}{\partial x_\alpha} \tag{12}$$

New tensors, in respect to linear orthogonal transformations, may be formed from tensors according to these rules of operation.

*Symmetry Properties of Tensors.* Tensors are called symmetrical or skew-symmetrical in respect to two of their indices, $\mu$ and $\nu$, if both the components which result from interchanging the indices $\mu$ and $\nu$ are equal to each other or equal with opposite signs.

Condition for symmetry: $A_{\mu\nu\rho} = A_{\nu\mu\rho}$.

Condition for skew-symmetry: $A_{\mu\nu\rho} = -A_{\nu\mu\rho}$.

*Theorem.* The character of symmetry or skew-symmetry exists independently of the choice of co-ordinates, and in this lies its importance. The proof follows from the equation defining tensors.

*Special Tensors.*

I. The quantities $\delta_{\rho\sigma}$ (4) are tensor components (fundamental tensor).

*Proof.* If in the right-hand side of the equation of transformation $A'_{\mu\nu} = b_{\mu\alpha}b_{\nu\beta}A_{\alpha\beta}$, we substitute for $A_{\alpha\beta}$ the quantities $\delta_{\alpha\beta}$ (which are equal to 1 or 0 according as $\alpha = \beta$ or $\alpha \neq \beta$), we get

$$A'_{\mu\nu} = b_{\mu\alpha}b_{\nu\alpha} = \delta_{\mu\nu}.$$

The justification for the last sign of equality becomes evident if one applies (4) to the inverse substitution (5).

II. There is a tensor $(\delta_{\mu\nu\rho} \ldots)$ skew-symmetrical with respect to all pairs of indices, whose rank is equal to the number of dimensions, $n$, and whose components are equal to $+1$ or $-1$ according as $\mu\nu\rho \ldots$ is an even or odd permutation of 123 . . .

The proof follows with the aid of the theorem proved above $|b_{\rho\sigma}| = 1$.

These few simple theorems form the apparatus from the theory of invariants for building the equations of pre-relativity physics and the theory of special relativity.

We have seen that in pre-relativity physics, in order to specify relations in space, a body of reference, or a space of reference, is required, and, in addition, a Cartesian system of co-ordinates. We can fuse both these concepts into a single one by thinking of a Cartesian system of co-ordinates as a cubical frame-work formed of rods each of unit length. The co-ordinates of the lattice points of this frame are integral numbers. It follows from the fundamental relation

$$s^2 = \Delta x_1^2 + \Delta x_2^2 + \Delta x_3^2 \tag{13}$$

that the members of such a space-lattice are all of unit

length. To specify relations in time, we require in addition a standard clock placed, say, at the origin of our Cartesian system of co-ordinates or frame of reference. If an event takes place anywhere we can assign to it three co-ordinates, $x_\nu$, and a time $t$, as soon as we have specified the time of the clock at the origin which is simultaneous with the event. We therefore give (hypothetically) an objective significance to the statement of the simultaneity of distant events, while previously we have been concerned only with the simultaneity of two experiences of an individual. The time so specified is at all events independent of the position of the system of co-ordinates in our space of reference, and is therefore an invariant with respect to the transformation (3).

It is postulated that the system of equations expressing the laws of pre-relativity physics is co-variant with respect to the transformation (3), as are the relations of Euclidean geometry. The isotropy and homogeneity of space is expressed in this way.* We shall now consider some of the more important equations of physics from this point of view.

The equations of motion of a material particle are

$$m \frac{d^2 x_\nu}{dt^2} = X_\nu \tag{14}$$

$(dx_\nu)$ is a vector; $dt$, and therefore also $\frac{1}{dt}$, an invariant;

* The laws of physics could be expressed, even in case there were a preferred direction in space, in such a way as to be co-variant with respect to the transformation (3); but such an expression would in this case be unsuitable. If there were a preferred direction in space it would simplify the description of natural phenomena to orient the system of co-ordinates in a definite way with respect to this direction. But if, on the other hand, there is no unique direction in space it is not logical to formulate the laws of nature in such a way as to conceal the equivalence of systems of co-ordinates that are oriented differently. We shall meet with this point of view again in the theories of special and general relativity.

thus $\left(\frac{dx_\nu}{dt}\right)$ is a vector; in the same way it may be shown that $\left(\frac{d^2x_\nu}{dt^2}\right)$ is a vector. In general, the operation of differentiation with respect to time does not alter the tensor character. Since $m$ is an invariant (tensor of rank 0), $\left(m\frac{d^2x_\nu}{dt^2}\right)$ is a vector, or tensor of rank 1 (by the theorem of the multiplication of tensors). If the force $(X_\nu)$ has a vector character, the same holds for the difference $\left(m\frac{d^2x_\nu}{dt^2} - X_\nu\right)$. These equations of motion are therefore valid in every other system of Cartesian co-ordinates in the space of reference. In the case where the forces are conservative we can easily recognize the vector character of $(X_\nu)$. For a potential energy, $\Phi$, exists, which depends only upon the mutual distances of the particles, and is therefore an invariant. The vector character of the force, $X_\nu = -\frac{\partial\Phi}{\partial x_\nu}$, is then a consequence of our general theorem about the derivative of a tensor of rank 0.

Multiplying by the velocity, a tensor of rank 1, we obtain the tensor equation

$$\left(m\frac{d^2x_\nu}{dt^2} - X_\nu\right)\frac{dx_\mu}{dt} = 0.$$

By contraction and multiplication by the scalar $dt$ we obtain the equation of kinetic energy

$$d\left(\frac{mq^2}{2}\right) = X_\nu dx_\nu.$$

If $\xi_\nu$ denotes the difference of the co-ordinates of the material particle and a point fixed in space, then the $\xi_\nu$ have vector character. We evidently have

$\frac{d^2x_\nu}{dt^2} = \frac{d^2\xi_\nu}{dt^2}$, so that the equations of motion of the particle may be written

$$m\frac{d^2\xi_\nu}{dt^2} - X_\nu = 0.$$

Multiplying this equation by $\xi_\mu$ we obtain a tensor equation

$$\left(m\frac{d^2\xi_\nu}{dt^2} - X_\nu\right)\xi_\mu = 0.$$

Contracting the tensor on the left and taking the time average we obtain the virial theorem, which we shall not consider further. By interchanging the indices and subsequent subtraction, we obtain, after a simple transformation, the theorem of moments,

$$\frac{d}{dt}\left[m\left(\xi_\mu\frac{d\xi_\nu}{dt} - \xi_\nu\frac{d\xi_\mu}{dt}\right)\right] = \xi_\mu X_\nu - \xi_\nu X_\mu \tag{15}$$

It is evident in this way that the moment of a vector is not a vector but a tensor. On account of their skew-symmetrical character there are not nine, but only three independent equations of this system. The possibility of replacing skew-symmetrical tensors of the second rank in space of three dimensions by vectors depends upon the formation of the vector

$$A_\mu = \frac{1}{2}A_{\sigma\tau}\delta_{\sigma\tau\mu}.$$

If we multiply the skew-symmetrical tensor of rank 2 by the special skew-symmetrical tensor $\delta$ introduced above, and contract twice, a vector results whose components are numerically equal to those of the tensor. These are the so-called axial vectors which transform differently, from a right-handed system to a left-handed system, from the $\Delta x_\nu$. There is a gain in picturesqueness in regarding a skew-symmetrical tensor of rank 2 as a vector in space of three dimensions, but it does not represent the exact nature of the corresponding quantity so well as considering it a tensor.

We consider next the equations of motion of a continuous medium. Let $\rho$ be the density, $u_\nu$ the velocity components considered as functions of the co-ordinates and the time, $X_\nu$ the volume forces per unit of mass, and $p_{\nu\sigma}$ the stresses upon a surface perpendicular to the $\sigma$-axis in the direction of increasing $x_\nu$. Then the equations of motion area, by Newton's law,

$$\rho \frac{du_\nu}{dt} = - \frac{\partial p_{\nu\sigma}}{\partial x_\sigma} + \rho X_\nu$$

in which $\frac{du_\nu}{dt}$ is the acceleration of the particle which at time $t$ has the co-ordinates $x_\nu$. If we express this acceleration by partial differential coefficients, we obtain, after dividing by $\rho$,

$$\frac{\partial u_\nu}{\partial t} + \frac{\partial u_\nu}{\partial x_\sigma} u_\sigma = - \frac{1}{\rho} \frac{\partial p_{\nu\sigma}}{\partial x_\sigma} + X_\nu \tag{16}$$

We must show that this equation holds independently of the special choice of the Cartesian system of co-ordinates. $(u_\nu)$ is a vector, and therefore $\frac{\partial u_\nu}{\partial t}$ is also a vector. $\frac{\partial u_\nu}{\partial x_\sigma}$ is a tensor of rank 2, $\frac{\partial u_\nu}{\partial x_\sigma} u_\tau$ is a tensor of rank 3. The second term on the left results from contraction in the indices $\sigma, \tau$. The vector character of the second term on the right is obvious. In order that the first term on the right may also be a vector it is necessary for $p_{\nu\sigma}$ to be a tensor. Then by differentiation and contraction $\frac{\partial p_{\nu\sigma}}{\partial x_\sigma}$ results, and is therefore a vector, as it also is after multiplication by the reciprocal scalar $\frac{1}{\rho}\cdot$ That $p_{\nu\sigma}$ is a tensor, and therefore transforms according to the equation

$$p'_{\mu\nu} = b_{\mu\alpha} b_{\nu\beta} p_{\alpha\beta},$$

is proved in mechanics by integrating this equation over an infinitely small tetrahedron. It is also proved there, by application of the theorem of moments to an infinitely small parallelepipedon, that $p_{\nu\sigma} = p_{\sigma\nu}$, and hence that the tensor of the stress is a symmetrical tensor. From what has been said it follows that, with the aid of the rules given above, the equation is co-variant with respect to orthogonal transformations in space (rotational transformations); and the rules according to which the quantities in the equation must be transformed in order that the equation may be co-variant also become evident.

The co-variance of the equation of continuity,

$$\frac{\partial \rho}{\partial t} + \frac{\partial (\rho u_\nu)}{\partial x_\nu} = 0 \tag{17}$$

requires, from the foregoing, no particular discussion.

We shall also test for co-variance the equations which express the dependence of the stress components upon the properties of the matter, and set up these equations for the case of a compressible viscous fluid with the aid of the conditions of co-variance. If we neglect the viscosity, the pressure, $p$, will be a scalar, and will depend only upon the density and the temperature of the fluid.

The contribution to the stress tensor is then evidently

$$p\delta_{\mu\nu}$$

in which $\delta_{\mu\nu}$ is the special symmetrical tensor. This term will also be present in the case of a viscous fluid. But in this case there will also be pressure terms, which depend upon the space derivatives of the $u_\nu$. We shall assume that this dependence is a linear one. Since these terms must be symmetrical tensors, the only ones which enter will be

$$\alpha\left(\frac{\partial u_\mu}{\partial x_\nu} + \frac{\partial u_\nu}{\partial x_\mu}\right) + \beta\delta_{\mu\nu}\frac{\partial u_\alpha}{\partial x_\alpha}$$

$\left(\text{for } \frac{\partial u_\alpha}{\partial x_\alpha} \text{ is a scalar}\right)$. For physical reasons (no slipping) it is assumed that for symmetrical dilatations in all directions, i.e. when

$$\frac{\partial u_1}{\partial x_1} = \frac{\partial u_2}{\partial x_2} = \frac{\partial u_3}{\partial x_3}; \frac{\partial u_1}{\partial x_2}, \text{ etc.}, = 0,$$

there are no frictional forces present, from which it follows that $\beta = -\frac{2}{3}\alpha$. If only $\frac{\partial u_1}{\partial x_3}$ is different from zero, let $p_{31} = -\eta\frac{\partial u_1}{\partial x_3}$, by which $\alpha$ is determined. We then obtain for the complete stress tensor,

$$p_{\mu\nu} = p\delta_{\mu\nu} - \eta\left[\left(\frac{\partial u_\mu}{\partial x_\nu} + \frac{\partial u_\nu}{\partial x_\mu}\right) - \frac{2}{3}\left(\frac{\partial u_1}{\partial x_1} + \frac{\partial u_2}{\partial x_2} + \frac{\partial u_3}{\partial x_3}\right)\delta_{\mu\nu}\right] \tag{18}$$

The heuristic value of the theory of invariants, which arises from the isotropy of space (equivalence of all directions), becomes evident from this example.

We consider, finally, Maxwell's equations in the form which are the foundation of the electron theory of Lorentz.

$$(19)\quad \left\{\begin{array}{l} \dfrac{\partial h_3}{\partial x_2} - \dfrac{\partial h_2}{\partial x_3} = \dfrac{1}{c}\dfrac{\partial e_1}{\partial t} + \dfrac{1}{c} i_1 \\ \dfrac{\partial h_1}{\partial x_3} - \dfrac{\partial h_3}{\partial x_1} = \dfrac{1}{c}\dfrac{\partial e_2}{\partial t} + \dfrac{1}{c} i_2 \\ \cdot \quad \cdot \quad \cdot \quad \cdot \quad \cdot \\ \dfrac{\partial e_1}{\partial x_1} + \dfrac{\partial e_2}{\partial x_2} + \dfrac{\partial e_3}{\partial x_3} = \rho \end{array}\right.$$

$$(20)\quad \left\{\begin{array}{l} \dfrac{\partial e_3}{\partial x_2} - \dfrac{\partial e_2}{\partial x_3} = -\dfrac{1}{c}\dfrac{\partial h_1}{\partial t} \\ \dfrac{\partial e_1}{\partial x_3} - \dfrac{\partial e_3}{\partial x_1} = -\dfrac{1}{c}\dfrac{\partial h_2}{\partial t} \\ \cdot \quad \cdot \quad \cdot \quad \cdot \\ \dfrac{\partial h_1}{\partial x_1} + \dfrac{\partial h_2}{\partial x_2} + \dfrac{\partial h_3}{\partial x_3} = 0 \end{array}\right.$$

**i** is a vector, because the current density is defined as the density of electricity multiplied by the vector velocity of the electricity. According to the first three equations it is evident that **e** is also to be regarded as a vector. Then **h** cannot be regarded as a vector.* The equations may, however, easily be interpreted if **h** is regarded as a skew-symmetrical tensor of the second rank. Accordingly, we write $h_{23}$, $h_{31}$, $h_{12}$, in place of $h_1$, $h_2$, $h_3$ respectively. Paying attention to the skew-symmetry of $h_{\mu\nu}$, the first three equations of (19) and (20) may be written in the form

$$(19a)\quad \frac{\partial h_{\mu\nu}}{\partial x_\nu} = \frac{1}{c}\frac{\partial e_\mu}{\partial t} + \frac{1}{c} i_\mu$$

$$(20a)\quad \frac{\partial e_\mu}{\partial x_\nu} - \frac{\partial e_\nu}{\partial x_\mu} = +\frac{1}{c}\frac{\partial h_{\mu\nu}}{\partial t}$$

* These considerations will make the reader familiar with tensor operations without the special difficulties of the four-dimensional treatment; corresponding considerations in the theory of special relativity (Minkowski's interpretation of the field) will then offer fewer difficulties.

In contrast to **e**, **h** appears as a quantity which has the same type of symmetry as an angular velocity. The divergence equations then take the form

(19b)
$$\frac{\partial e_\nu}{\partial x_\nu} = \rho$$

(20b)
$$\frac{\partial h_{\mu\nu}}{\partial x_\rho} + \frac{\partial h_{\nu\rho}}{\partial x_\mu} + \frac{\partial h_{\rho\mu}}{\partial x_\nu} = 0$$

The last equation is a skew-symmetrical tensor equation of the third rank (the skew-symmetry of the left-hand side with respect to every pair of indices may easily be proved, if attention is paid to the skew-symmetry of $h_{\mu\nu}$). This notation is more natural than the usual one, because, in contrast to the latter, it is applicable to Cartesian left-handed systems as well as to right-handed systems without change of sign.

# THE THEORY OF SPECIAL RELATIVITY

THE previous considerations concerning the configuration of rigid bodies have been founded, irrespective of the assumption as to the validity of the Euclidean geometry, upon the hypothesis that all directions in space, or all configurations of Cartesian systems of co-ordinates, are physically equivalent. We may express this as the "principle of relativity with respect to direction," and it has been shown how equations (laws of nature) may be found, in accord with this principle, by the aid of the calculus of tensors. We now inquire whether there is a relativity with respect to the state of motion of the space of reference; in other words, whether there are spaces of reference in motion relatively to each other which are physically equivalent. From the standpoint of mechanics it appears that equivalent spaces of reference do exist. For experiments upon the earth tell us nothing of the fact that we are moving about the sun with a velocity of approximately 30 kilometres a second. On the other hand, this physical equivalence does not seem to hold for spaces of reference in arbitrary motion; for mechanical effects do not seem to be subject to the same laws in a jolting railway train as in one moving with uniform velocity; the rotation of the earth must be considered in writing down the equations of motion relatively to the earth. It appears, therefore, as if there were Cartesian systems of co-ordinates, the so-called inertial systems, with reference to which the laws of mechanics (more generally the laws of physics) are expressed in the simplest form. We may surmise the validity of the following proposition: If $K$ is an inertial system, then every other system $K'$ which moves uniformly and without rotation relatively to $K$, is also an inertial system; the laws of nature are in concordance for all inertial systems. This statement we shall call the "principle of special relativity." We shall draw certain conclusions from this principle of "relativity of translation" just as we have already done for relativity of direction.

In order to be able to do this, we must first solve the following problem. If we are given the Cartesian co-ordinates, $x_\nu$, and the time $t$, of an event relatively to one inertial system, $K$, how can we calculate the co-ordinates, $x'_\nu$, and the time, $t'$, of the same event relatively to an inertial system $K'$ which moves with uniform translation relatively to $K$? In the pre-relativity physics this problem was solved by making unconsciously two hypotheses:—

1. Time is absolute; the time of an event, $t'$, relatively to $K'$ is the same as the time relatively to $K$. If instantaneous signals could be sent to a distance, and if one knew that the state of motion of a clock had no influence on its rate, then this assumption would be physically validated. For then clocks, similar to one another, and regulated alike, could be distributed over the systems $K$ and $K'$, at rest relatively to them, and their indications would be independent of the state of motion of the systems; the time of an event would then be given by the clock in its immediate neighbourhood.

2. Length is absolute; if an interval, at rest relatively to $K$, has a length $s$, then it has the same length $s$, relatively to a system $K'$ which is in motion relatively to $K$.

If the axes of $K$ and $K'$ are parallel to each other, a simple calculation based on these two assumptions, gives the equations of transformation

$$(21) \qquad \begin{cases} x'_\nu = x_\nu - a_\nu - b_\nu t \\ t' = t - b \end{cases}$$

This transformation is known as the "Galilean Transformation." Differentiating twice by the time, we get

$$\frac{d^2x'_\nu}{dt^2} = \frac{d^2x_\nu}{dt^2}.$$

Further, it follows that for two simultaneous events,

$$x'^{(1)}_\nu - x'^{(2)}_\nu = x^{(1)}_\nu - x^{(2)}_\nu.$$

The invariance of the distance between the two points results from squaring and adding. From this easily follows the co-variance of Newton's equations of motion with respect to the Galilean transformation (21). Hence it follows that classical mechanics is in accord with the principle of special relativity if the two hypotheses respecting scales and clocks are made.

But this attempt to found relativity of translation upon the Galilean transformation fails when applied to electromagnetic phenomena. The Maxwell-Lorentz electromagnetic equations are not co-variant with respect to the Galilean transformation. In particular, we note, by (21), that a ray of light which referred to $K$ has a velocity $c$, has a different velocity referred to $K'$, depending upon its direction. The space of reference of $K$ is therefore distinguished, with respect to its physical properties, from all spaces of reference which are in motion relatively to it (quiescent ether). But all experiments have shown that electro-magnetic and optical phenomena, relatively to the earth as the body of reference, are not influenced by the translational velocity of the earth. The most important of these experiments are those of Michelson and Morley, which I shall assume are known. The validity of the principle of special relativity also with respect to electromagnetic phenomena can therefore hardly be doubted.

On the other hand, the Maxwell-Lorentz equations have proved their validity in the treatment of optical problems in moving bodies. No other theory has satisfactorily explained the facts of aberration, the propagation of light in moving bodies (Fizeau), and phenomena observed in double stars (De Sitter). The consequence of the Maxwell-Lorentz equations that in a vacuum light is propagated with the velocity $c$, at least with respect to a definite inertial system $K$, must therefore be regarded as proved. According to the principle of special relativity, we must also assume the truth of this principle for every other inertial system.

Before we draw any conclusions from these two principles we must first review the physical significance of the concepts "time" and "velocity." It follows from what has gone before, that co-ordinates with respect to an inertial system are physically defined by means of measurements and constructions with the aid of rigid bodies. In order to measure time, we have supposed a clock, $U$, present somewhere, at rest relatively to $K$. But we cannot fix the time, by means of this clock, of an event whose distance from the clock is not negligible; for there are no "instantaneous signals" that we can use in order to compare the time of the event with that of the clock. In order to complete the definition of time we may employ the principle of the constancy of the velocity of light in a vacuum. Let us suppose that we place similar clocks at points of the system $K$, at rest relatively to it, and regulated according to the following scheme. A ray of light is sent out from one of the clocks, $U_m$, at the instant when it indicates the time $t_m$, and travels through a vacuum a distance $r_{mn}$, to the clock $U_n$; at the instant when this ray meets the clock $U_n$ the latter is set to indicate the time $t_n = t_m + \frac{r_{mn}}{c}$.* The principle of the constancy of the velocity of light then states that this adjustment of the clocks will not lead to contradictions. With clocks so adjusted, we can assign the time to events which take place near any one of them. It is essential to note that this definition of time relates only to the inertial system $K$, since we have used a system of clocks at rest relatively to $K$. The assumption which was made in the pre-relativity physics of the absolute character of time (i.e. the independence of time of the choice of the inertial system) does not follow at all from this definition.

The theory of relativity is often criticized for giving, without justification, a central theoretical rôle to the propagation of light, in that it founds the concept of time upon the law of propagation of light. The situation, however, is somewhat as follows. In order to give physical significance to the concept of time, processes of some kind are required which enable relations to be established between different places. It is immaterial what kind of processes one chooses for such a definition of time. It is advantageous, however, for the theory, to choose only those processes concerning which we know something

* Strictly speaking, it would be more correct to define simultaneity first somewhat as follows: two events taking place at the points $A$ and $B$ of the system $K$ are simultaneous if they appear at the same instant when observed from the middle point, $M$, of the interval $AB$. Time is then defined as the ensemble of the indications of similar clocks, at rest relatively to $K$, which register the same time simultaneously.

certain. This holds for the propagation of light *in vacuo* in a higher degree than for any other process which could be considered, thanks to the investigations of Maxwell and H. A. Lorentz.

From all of these considerations, space and time data have a physically real, and not a mere fictitious, significance; in particular this holds for all the relations in which co-ordinates and time enter, e.g. the relations (21). There is, therefore, sense in asking whether those equations are true or not, as well as in asking what the true equations of transformation are by which we pass from one inertial system $K$ to another, $K'$, moving relatively to it. It may be shown that this is uniquely settled by means of the principle of the constancy of the velocity of light and the principle of special relativity.

To this end we think of space and time physically defined with respect to two inertial systems, $K$ and $K'$, in the way that has been shown. Further, let a ray of light pass from one point $P_1$ to another point $P_2$ of $K$ through a vacuum. If $r$ is the measured distance between the two points, then the propagation of light must satisfy the equation

$$r = c \, . \, \Delta t.$$

If we square this equation, and express $r^2$ by the differences of the co-ordinates, $\Delta x_\nu$, in place of this equation we can write

$$\sum (\Delta x_\nu)^2 - c^2 \Delta t^2 = 0 \tag{22}$$

This equation formulates the principle of the constancy of the velocity of light relatively to $K$. It must hold whatever may be the motion of the source which emits the ray of light.

The same propagation of light may also be considered relatively to $K'$, in which case also the principle of the constancy of the velocity of light must be satisfied. Therefore, with respect to $K'$, we have the equation

$$\sum (\Delta x'_\nu)^2 - c^2 \Delta t'^2 = 0 \tag{22a}$$

Equations (22a) and (22) must be mutually consistent with each other with respect to the transformation which transforms from $K$ to $K'$. A transformation which effects this we shall call a "Lorentz transformation."

Before considering these transformations in detail we shall make a few general remarks about space and time. In the pre-relativity physics space and time were separate entities. Specifications of time were independent of the choice of the space of reference. The Newtonian mechanics was relative with respect to the space of reference, so that, e.g. the statement that two non-simultaneous events happened at the same place had no objective meaning (that is, independent of the space of reference). But this relativity had no rôle in building up the theory. One spoke of points of space, as of instants of time, as if they were absolute realities. It was not observed that the true element of the space-time specification was the event specified by the four numbers $x_1$, $x_2$, $x_3$, $t$. The conception of something happening was always that of a four-dimensional continuum; but the recognition of this was obscured by the absolute character of the pre-relativity time. Upon giving up the hypothesis of the absolute character of time, particularly that of simultaneity, the four-dimensionality of the time-space concept was immediately recognized. It is neither the point in space, nor the instant in time, at which something happens that has physical reality, but only the event itself. There is no absolute (independent of the space of reference) relation in space, and no absolute relation in time between two events, but there is an absolute (independent of the space of reference) relation in space and time, as will appear in the sequel. The circumstance that there is no objective rational division of the four-dimensional continuum into a three-dimensional space and a one-dimensional time continuum indicates that the laws of nature will assume a form which is logically most satisfactory when expressed as laws in the four-dimensional space-time continuum. Upon this depends the great advance in method which the theory of relativity owes to Minkowski. Considered from this standpoint, we must regard $x_1$, $x_2$, $x_3$, $t$ as the four co-ordinates of an event in the four-dimensional continuum. We have far less success in picturing to ourselves relations in this four-dimensional continuum than in the three-dimensional Euclidean continuum; but it must be emphasized that even in the Euclidean three-dimensional geometry its concepts and relations are only of an abstract nature in our minds, and are not at all identical with the images we form visually and through our sense of touch. The non-divisibility of the four-dimensional continuum of events does not at all, however, involve the equivalence of the space co-ordinates with the time co-ordinate. On the contrary, we must remember that the time co-ordinate is defined physically wholly differently from the space co-ordinates. The relations (22) and (22a) which when equated define the Lorentz transformation show, further, a difference in the rôle of the time co-ordinate from that of the space co-ordinates; for the term $\Delta t^2$ has the opposite sign to the space terms, $\Delta x_1^2$, $\Delta x_2^2$, $\Delta x_3^2$.

Before we analyse further the conditions which define the Lorentz transformation, we shall introduce the light-time, $l = ct$, in place of the time, $t$, in order that the constant $c$ shall not enter explicitly into the formulas to be developed later. Then the Lorentz transformation is defined in such a way that, first, it makes the equation

$$\Delta x_1^2 + \Delta x_2^2 + \Delta x_3^2 - \Delta l^2 = 0 \tag{22b}$$

a co-variant equation, that is, an equation which is satisfied with respect to every inertial system if it is satisfied in the inertial system to which we refer the two given events (emission and reception of the ray of light). Finally, with Minkowski, we introduce in place of the real time co-ordinate $l = ct$, the imaginary time co-ordinate

$$x_4 = il = ict \ (\sqrt{-1} = i).$$

Then the equation defining the propagation of light, which must be co-variant with respect to the Lorentz transformation, becomes

$$\sum_{(4)} \Delta x_\nu^2 = \Delta x_1^2 + \Delta x_2^2 + \Delta x_3^2 + \Delta x_4^2 = 0 \tag{22c}$$

This condition is always satisfied* if we satisfy the more general condition that

$$s^2 = \Delta x_1^2 + \Delta x_2^2 + \Delta x_3^2 + \Delta x_4^2 \tag{23}$$

shall be an invariant with respect to the transformation. This condition is satisfied only by linear transformations, that is, transformations of the type

$$x'_\mu = a_\mu + b_{\mu\alpha} x_\alpha \tag{24}$$

in which the summation over the $\alpha$ is to be extended from $\alpha = 1$ to $\alpha = 4$. A glance at equations (23) (24) shows that the Lorentz transformation so defined is identical

* That this specialization lies in the nature of the case will be evident later.

with the translational and rotational transformations of the Euclidean geometry, if we disregard the number of dimensions and the relations of reality. We can also conclude that the coefficients $b_{\mu\alpha}$ must satisfy the conditions

$$b_{\mu\alpha}b_{\nu\alpha} = \delta_{\mu\nu} = b_{\alpha\mu}b_{\alpha\nu} \tag{25}$$

Since the ratios of the $x_\nu$ are real, it follows that all the $a_\mu$ and the $b_{\mu\alpha}$ are real, except $a_4$, $b_{41}$, $b_{42}$, $b_{43}$, $b_{14}$, $b_{24}$, and $b_{34}$, which are purely imaginary.

*Special Lorentz Transformation.* We obtain the simplest transformations of the type of (24) and (25) if only two of the co-ordinates are to be transformed, and if all the $a_\mu$, which merely determine the new origin, vanish. We obtain then for the indices 1 and 2, on account of the three independent conditions which the relations (25) furnish,

$$\left\{\begin{aligned} x'_1 &= x_1 \cos\phi - x_2 \sin\phi \\ x'_2 &= x_1 \sin\phi + x_2 \cos\phi \\ x'_3 &= x_3 \\ x'_4 &= x_4 \end{aligned}\right. \tag{26}$$

This is a simple rotation in space of the (space) co-ordinate system about the $x_3$-axis. We see that the rotational transformation in space (without the time transformation) which we studied before is contained in the Lorentz transformation as a special case. For the indices 1 and 4 we obtain, in an analogous manner,

$$\left\{\begin{aligned} x'_1 &= x_1 \cos\psi - x_4 \sin\psi \\ x'_4 &= x_1 \sin\psi + x_4 \cos\psi \\ x'_2 &= x_2 \\ x'_3 &= x_3 \end{aligned}\right. \tag{26a}$$

On account of the relations of reality $\psi$ must be taken as imaginary. To interpret these equations physically, we introduce the real light-time $l$ and the velocity $v$ of $K'$ relatively to $K$, instead of the imaginary angle $\psi$. We have, first,

$$\begin{aligned} x'_1 &= x_1 \cos\psi - il \sin\psi \\ l' &= -ix_1 \sin\psi + l \cos\psi. \end{aligned}$$

Since for the origin of $K'$, i.e., for $x'_1 = 0$, we must have $x_1 = vl$, it follows from the first of these equations that

$$v = i \tan\psi \tag{27}$$

and also

$$\left\{ \begin{aligned} \sin\psi &= \frac{-iv}{\sqrt{1-v^2}} \\ \cos\psi &= \frac{1}{\sqrt{1-v^2}} \end{aligned} \right. \tag{28}$$

so that we obtain

$$\left\{ \begin{aligned} x'_1 &= \frac{x_1 - vl}{\sqrt{1-v^2}} \\ l' &= \frac{l - vx_1}{\sqrt{1-v^2}} \\ x'_2 &= x_2 \\ x'_3 &= x_3 \end{aligned} \right. \tag{29}$$

These equations form the well-known special Lorentz transformation, which in the general theory represents a rotation, through an imaginary angle, of the four-dimensional system of co-ordinates. If we introduce the ordinary time $t$, in place of the light-time $l$, then in (29) we must replace $l$ by $ct$ and $v$ by $\frac{v}{c}$.

We must now fill in a gap. From the principle of the constancy of the velocity of light it follows that the equation

$$\sum \Delta x_\nu^2 = 0$$

has a significance which is independent of the choice of the inertial system; but the invariance of the quantity $\sum \Delta x_\nu^2$ does not at all follow from this. This quantity might be transformed with a factor. This depends upon the fact that the right-hand side of (29) might be multiplied by a factor $\lambda$, which may depend on $v$. But the principle of relativity does not permit this factor to be different from 1, as we shall now show. Let us assume that we have a rigid circular cylinder moving in the direction of its axis. If its radius, measured at rest with a unit measuring rod is equal to $R_0$, its radius $R$ in motion, might be different from $R_0$, since the theory of relativity does not make the assumption that the shape of bodies with respect to a space of reference is independent of their motion relatively to this space of reference. But all directions in space must be equivalent to each other. $R$ may therefore depend upon the magnitude $q$ of the velocity, but not upon its direction; $R$ must therefore be an even function of $q$. If the cylinder is at rest relatively to $K'$ the equation of its lateral surface is

$$x'^2 + y'^2 = R_0^2.$$

If we write the last two equations of (29) more generally

$$\begin{aligned} x'_2 &= \lambda x_2 \\ x'_3 &= \lambda x_3 \end{aligned}$$

then the lateral surface of the cylinder referred to $K$ satisfies the equation

$$x^2 + y^2 = \frac{R_0^2}{\lambda^2}.$$

The factor $\lambda$ therefore measures the lateral contraction of the cylinder, and can thus, from the above, be only an even function of $v$.

If we introduce a third system of co-ordinates, $K''$, which moves relatively to $K'$ with velocity $v$ in the direction of the negative $x$-axis of $K$, we obtain, by applying (29) twice,

$$x''_1 = \lambda(v)\lambda(-v)x_1$$

$$\cdot \quad \cdot \quad \cdot \quad \cdot$$

$$\cdot \quad \cdot \quad \cdot \quad \cdot$$

$$l'' = \lambda(v)\lambda(-v)l.$$

Now, since $\lambda(v)$ must be equal to $\lambda(-v)$, and since we assume that we use the same measuring rods in all the systems, it follows that the transformation of $K''$ to $K$ must be the identical transformation (since the possibility $\lambda = -1$ does not need to be considered). It is essential for these considerations to assume that the behaviour of the measuring rods does not depend upon the history of their previous motion.

*Moving Measuring Rods and Clocks.* At the definite $K$ time, $l = 0$, the position of the points given by the integers $x'_1 = n$, is with respect to $K$, given by $x_1 = n\sqrt{1 - v^2}$; this follows from the first of equations (29) and expresses the Lorentz contraction. A clock at rest at the origin $x_1 = 0$ of $K$, whose beats are characterized by $l = n$, will, when observed from $K'$, have beats characterized by

$$l' = \frac{n}{\sqrt{1 - v^2}};$$

this follows from the second of equations (29) and shows that the clock goes slower than if it were at rest relatively to $K'$. These two consequences, which hold, *mutatis mutandis*, for every system of reference, form the physical content, free from convention, of the Lorentz transformation.

*Addition Theorem for Velocities.* If we combine two special Lorentz transformations with the relative velocities $v_1$ and $v_2$, then the velocity of the single Lorentz transformation which takes the place of the two separate ones is, according to (27), given by

$$v_{12} = i \tan(\psi_1 + \psi_2) = i \frac{\tan\psi_1 + \tan\psi_2}{1 - \tan\psi_1 \tan\psi_2} = \frac{v_1 + v_2}{1 + v_1 v_2}. \tag{30}$$

*General Statements about the Lorentz Transformation and its Theory of Invariants.* The whole theory of invariants of the special theory of relativity depends upon the invariant $s^2$ (23). Formally, it has the same rôle in the four-dimensional space-time continuum as the invariant $\Delta x_1^2 + \Delta x_2^2 + \Delta x_3^2$ in the Euclidean geometry and in the pre-relativity physics. The latter quantity is not an invariant with respect to all the Lorentz transformations; the quantity $s^2$ of equation (23) assumes the rôle of this invariant. With respect to an arbitrary inertial system, $s^2$ may be determined by measurements; with a given unit of measure it is a completely determinate quantity, associated with an arbitrary pair of events.

The invariant $s^2$ differs, disregarding the number of dimensions, from the corresponding invariant of the Euclidean geometry in the following points. In the Euclidean geometry $s^2$ is necessarily positive; it vanishes only when the two points concerned come together. On the other hand, from the vanishing of

$$s^2 = \sum \Delta x_\nu^2 = \Delta x_1^2 + \Delta x_2^2 + \Delta x_3^2 - \Delta t^2$$

it cannot be concluded that the two space-time points

fall together; the vanishing of this quantity $s^2$, is the invariant condition that the two space-time points can be connected by a light signal *in vacuo*. If $P$ is a point (event) represented in the four-dimensional space of the $x_1$, $x_2$, $x_3$, $l$, then all the "points" which can be connected to $P$ by means of a light signal lie upon the cone $s^2 = 0$ (compare Fig. 1, in which the dimension $x_3$ is suppressed). The "upper"

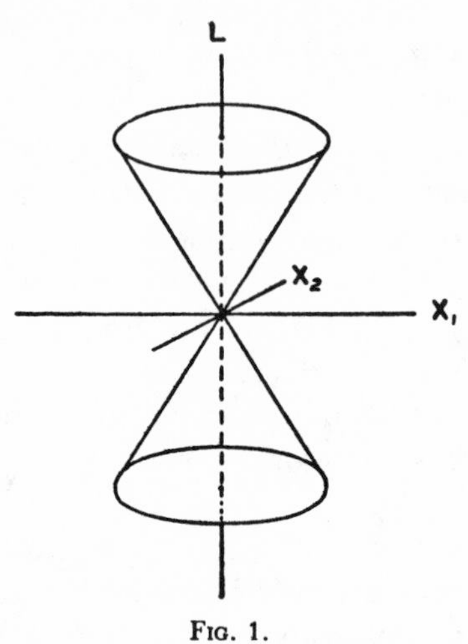


FIG. 1.

half of the cone may contain the "points" to which light signals can be sent from $P$; then the "lower" half of the cone will contain the "points" from which light signals can be sent to $P$. The points $P'$ enclosed by the conical surface furnish, with $P$, a negative $s^2$; $PP'$, as well as $P'P$ is then, according to Minkowski, time-like. Such intervals represent elements of possible paths of motion, the velocity being less than that of light.* In this case the $l$-axis

* That material velocities exceeding that of light are not possible, follows from the appearance of the radical $\sqrt{1 - v^2}$ in the special Lorentz transformation (29).

may be drawn in the direction of $PP'$ by suitably choosing the state of motion of the inertial system. If $P'$ lies outside of the "light-cone" then $PP'$ is space-like; in this case, by properly choosing the inertial system, $\Delta l$ can be made to vanish.

By the introduction of the imaginary time variable, $x_4 = il$, Minkowski has made the theory of invariants for the four-dimensional continuum of physical phenomena fully analogous to the theory of invariants for the three-dimensional continuum of Euclidean space. The theory of four-dimensional tensors of special relativity differs from the theory of tensors in three-dimensional space, therefore, only in the number of dimensions and the relations of reality.

A physical entity which is specified by four quantities, $A_\nu$, in an arbitrary inertial system of the $x_1$, $x_2$, $x_3$, $x_4$, is called a 4-vector, with the components $A_\nu$, if the $A_\nu$ correspond in their relations of reality and the properties of transformation to the $\Delta x_\nu$; it may be space-like or time-like. The sixteen quantities $A_{\mu\nu}$ then form the components of a tensor of the second rank, if they transform according to the scheme

$$A'_{\mu\nu} = b_{\mu\alpha} b_{\nu\beta} A_{\alpha\beta}.$$

It follows from this that the $A_{\mu\nu}$ behave, with respect to their properties of transformation and their properties of reality, as the products of the components, $U_\mu$, $V_\nu$, of two 4-vectors, $(U)$ and $(V)$. All the components are real except those which contain the index 4 once, those being purely imaginary. Tensors of the third and higher ranks may be defined in an analogous way. The operations of addition, subtraction, multiplication, contraction and differentiation for these tensors are wholly analogous to the corresponding operations for tensors in three-dimensional space.

Before we apply the tensor theory to the four-dimensional space-time continuum, we shall examine more particularly the skew-symmetrical tensors. The tensor of the second rank has, in general, 16 = 4.4 components. In the case of skew-symmetry the components with two equal indices vanish, and the components with unequal indices are equal and opposite in pairs. There exist, therefore, only six independent components, as is the case in the electromagnetic field. In fact, it will be shown when we consider Maxwell's equations that these may be looked upon as tensor equations, provided we regard the electromagnetic field as a skew-symmetrical tensor. Further, it is clear that a skew-symmetrical tensor of the third rank (skew-symmetrical in all pairs of indices) has only four independent components, since there are only four combinations of three different indices.

We now turn to Maxwell's equations (19a), (19b), (20a), (20b), and introduce the notation:*

$$\left\{\begin{array}{cccccc} \phi_{23} & \phi_{31} & \phi_{12} & \phi_{14} & \phi_{24} & \phi_{34} \\ h_{23} & h_{31} & h_{12} & -ie_x & -ie_y & -ie_z \end{array}\right. \tag{30a}$$

$$\left\{\begin{array}{cccc} \mathfrak{J}_1 & \mathfrak{J}_2 & \mathfrak{J}_3 & \mathfrak{J}_4 \\ \frac{1}{c}i_x & \frac{1}{c}i_y & \frac{1}{c}i_z & i\rho \end{array}\right. \tag{31}$$

with the convention that $\phi_{\mu\nu}$ shall be equal to $-\phi_{\nu\mu}$. Then Maxwell's equations may be combined into the forms

$$\frac{\partial \phi_{\mu\nu}}{\partial x_\nu} = \mathfrak{J}_\mu \tag{32}$$

$$\frac{\partial \phi_{\mu\nu}}{\partial x_\sigma} + \frac{\partial \phi_{\nu\sigma}}{\partial x_\mu} + \frac{\partial \phi_{\sigma\mu}}{\partial x_\nu} = 0 \tag{33}$$

* In order to avoid confusion from now on we shall use the three-dimensional space indices, $x, y, z$ instead of 1, 2, 3, and we shall reserve the numeral indices 1, 2, 3, 4 for the four-dimensional space-time continuum.

as one can easily verify by substituting from (30a) and (31). Equations (32) and (33) have a tensor character, and are therefore co-variant with respect to Lorentz transformations, if the $\phi_{\mu\nu}$ and the $\mathcal{J}_\mu$ have a tensor character, which we assume. Consequently, the laws for transforming these quantities from one to another allowable (inertial) system of co-ordinates are uniquely determined. The progress in method which electro-dynamics owes to the theory of special relativity lies principally in this, that the number of independent hypotheses is diminished. If we consider, for example, equations (19a) only from the standpoint of relativity of direction, as we have done above, we see that they have three logically independent terms. The way in which the electric intensity enters these equations appears to be wholly independent of the way in which the magnetic intensity enters them; it would not be surprising if instead of $\frac{\partial e_\mu}{\partial l}$, we had, say, $\frac{\partial^2 e_\mu}{\partial l^2}$, or if this term were absent. On the other hand, only two independent terms appear in equation (32). The electromagnetic field appears as a formal unit; the way in which the electric field enters this equation is determined by the way in which the magnetic field enters it. Besides the electromagnetic field, only the electric current density appears as an independent entity. This advance in method arises from the fact that the electric and magnetic fields lose their separate existences through the relativity of motion. A field which appears to be purely an electric field, judged from one system, has also magnetic field components when judged from another inertial system. When applied to an electromagnetic field, the general law of transformation furnishes, for the special case of the special Lorentz transformation, the equations

$$(34)\qquad \begin{cases} e'_x = e_x & h'_x = h_x \\ e'_y = \dfrac{e_y - vh_z}{\sqrt{1-v^2}} & h'_y = \dfrac{h_y + ve_z}{\sqrt{1-v^2}} \\ e'_z = \dfrac{e_z + vh_y}{\sqrt{1-v^2}} & h'_z = \dfrac{h_z - ve_y}{\sqrt{1-v^2}} \end{cases}$$

If there exists with respect to $K$ only a magnetic field, **h**, but no electric field, **e**, then with respect to $K'$ there exists an electric field **e**$'$ as well, which would act upon an electric particle at rest relatively to $K'$. An observer at rest relatively to $K$ would designate this force as the Biot-Savart force, or the Lorentz electromotive force. It therefore appears as if this electromotive force had become fused with the electric field intensity into a single entity.

In order to view this relation formally, let us consider the expression for the force acting upon unit volume of electricity,

$$(35)\qquad \mathbf{k} = \rho\mathbf{e} + \mathbf{i} \times \mathbf{h}$$

in which **i** is the vector velocity of electricity, with the velocity of light as the unit. If we introduce $J_\mu$ and $\phi_\mu$ according to (30a) and (31), we obtain for the first component the expression

$$\phi_{12}J_2 + \phi_{13}J_3 + \phi_{14}J_4.$$

Observing that $\phi_{11}$ vanishes on account of the skew-symmetry of the tensor $(\phi)$, the components of $k$ are given by the first three components of the four-dimensional vector

$$(36)\qquad K_\mu = \phi_{\mu\nu}J_\nu$$

and the fourth component is given by

$$(37)\qquad K_4 = \phi_{41}J_1 + \phi_{42}J_2 + \phi_{43}J_3 = i(e_x i_x + e_y i_y + e_z i_z) = i\lambda.$$

There is, therefore, a four-dimensional vector of force per unit volume, whose first three components, $k_1$, $k_2$, $k_3$, are the ponderomotive force components per unit volume, and whose fourth component is the rate of working of the field per unit volume, multiplied by $\sqrt{-1}$.

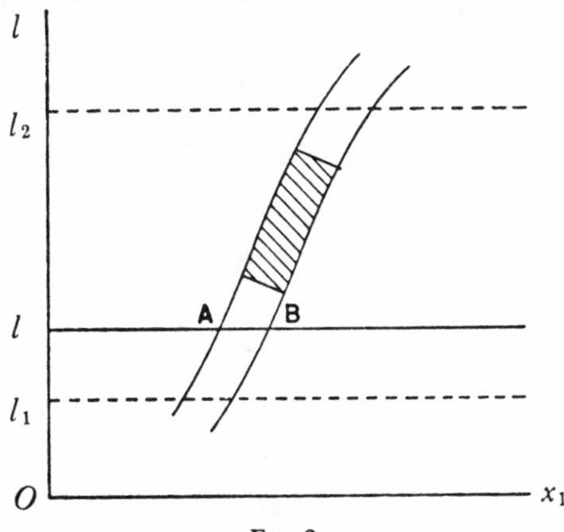


FIG. 2.

A comparison of (36) and (35) shows that the theory of relativity formally unites the ponderomotive force of the electric field, $\rho \mathbf{e}$, and the Biot-Savart or Lorentz force $\mathbf{i} \times \mathbf{h}$.

*Mass and Energy.* An important conclusion can be drawn from the existence and significance of the 4-vector $K_\mu$. Let us imagine a body upon which the electromagnetic field acts for a time. In the symbolic figure (Fig. 2) $Ox_1$ designates the $x_1$-axis, and is at the same time a substitute for the three space axes $Ox_1$, $Ox_2$, $Ox_3$; $Ol$ designates the real time axis. In this diagram a body of finite extent is represented, at a definite time $l$, by the interval AB; the whole space-time existence of the body is represented by a strip whose boundary is everywhere inclined less than 45° to the $l$-axis. Between the time sections, $l = l_1$ and $l = l_2$, but not extending to them, a portion of the strip is shaded. This represents the portion of the space-time manifold in which the electromagnetic field acts upon the body, or upon the electric charges contained in it, the action upon them being transmitted to the body. We shall now consider the changes which take place in the momentum and energy of the body as a result of this action.

We shall assume that the principles of momentum and energy are valid for the body. The change in momentum, $\Delta I_x$, $\Delta I_y$, $\Delta I_z$, and the change in energy, $\Delta E$, are then given by the expressions

$$\Delta I_x = \int_{l_0}^{l_1} dl \int k_x dxdydz = \frac{1}{i} \int K_1 dx_1 dx_2 dx_3 dx_4$$

$$\cdot \quad \cdot \quad \cdot \quad \cdot \quad \cdot \quad \cdot \quad \cdot \quad \cdot$$

$$\cdot \quad \cdot \quad \cdot \quad \cdot \quad \cdot \quad \cdot \quad \cdot \quad \cdot$$

$$\Delta E = \int_{l_0}^{l_1} dl \int \lambda dxdydz = \frac{1}{i} \int \frac{1}{i} K_4 dx_1 dx_2 dx_3 dx_4$$

Since the four-dimensional element of volume is an invariant, and $(K_1, K_2, K_3, K_4)$ forms a 4-vector, the four-dimensional integral extended over the shaded portion transforms as a 4-vector, as does also the integral between the limits $l_1$ and $l_2$, because the portion of the region which is not shaded contributes nothing to the integral. It follows, therefore, that $\Delta I_x$, $\Delta I_y$, $\Delta I_z$, $i\Delta E$ form a 4-vector. Since the quantities themselves may be presumed to transform in the same way as their increments, we infer that the aggregate of the four quantities

$$I_x, I_y, I_z, iE$$

has itself vector character; these quantities are referred

to an instantaneous condition of the body (e.g. at the time $l = l_1$).

This 4-vector may also be expressed in terms of the mass $m$, and the velocity of the body, considered as a material particle. To form this expression, we note first, that

$$(38) \quad -ds^2 = d\tau^2 = -(dx_1^2 + dx_2^2 + dx_3^2) - dx_4^2 = dl^2(1 - q^2)$$

is an invariant which refers to an infinitely short portion of the four-dimensional line which represents the motion of the material particle. The physical significance of the invariant $d\tau$ may easily be given. If the time axis is chosen in such a way that it has the direction of the line differential which we are considering, or, in other terms, if we transform the material particle to rest, we shall have $d\tau = dl$; this will therefore be measured by the light-seconds clock which is at the same place, and at rest relatively to the material particle. We therefore call $\tau$ the proper time of the material particle. As opposed to $dl$, $d\tau$ is therefore an invariant, and is practically equivalent to $dl$ for motions whose velocity is small compared to that of light. Hence we see that

$$(39) \quad u_\sigma = \frac{dx_\sigma}{d\tau}$$

has, just as the $dx_\nu$, the character of a vector; we shall designate $(u_\sigma)$ as the four-dimensional vector (in brief, 4-vector) of velocity. Its components satisfy, by (38), the condition

$$(40) \quad \sum u_\sigma^2 = -1.$$

We see that this 4-vector, whose components in the ordinary notation are

$$\frac{q_x}{\sqrt{1-q^2}}, \frac{q_y}{\sqrt{1-q^2}}, \frac{q_z}{\sqrt{1-q^2}}, \frac{i}{\sqrt{1-q^2}} \tag{41}$$

is the only 4-vector which can be formed from the velocity components of the material particle which are defined in three dimensions by

$$q_x = \frac{dx}{dl}, \; q_y = \frac{dy}{dl}, \; q_z = \frac{dz}{dl}.$$

We therefore see that

$$\left(m \frac{dx_\mu}{d\tau}\right) \tag{42}$$

must be that 4-vector which is to be equated to the 4-vector of momentum and energy whose existence we have proved above. By equating the components, we obtain, in three-dimensional notation,

$$\begin{cases} I_x = \dfrac{mq_x}{\sqrt{1-q^2}} \\ \cdot \quad \cdot \quad \cdot \\ \cdot \quad \cdot \quad \cdot \\ E = \dfrac{m}{\sqrt{1-q^2}} \end{cases} \tag{43}$$

We recognize, in fact, that these components of momentum agree with those of classical mechanics for velocities which are small compared to that of light. For large velocities the momentum increases more rapidly than linearly with the velocity, so as to become infinite on approaching the velocity of light.

If we apply the last of equations (43) to a material particle at rest ($q = 0$), we see that the energy, $E_0$, of a body at rest is equal to its mass. Had we chosen the second as our unit of time, we would have obtained

$$E_0 = mc^2. \tag{44}$$

Mass and energy are therefore essentially alike; they are only different expressions for the same thing. The mass of a body is not a constant; it varies with changes in its energy.* We see from the last of equations (43) that $E$ becomes infinite when $q$ approaches 1, the velocity of light. If we develop $E$ in powers of $q^2$, we obtain,

$$E = m + \frac{m}{2} q^2 + \frac{3}{8} mq^4 + \ldots \tag{45}$$

The second term of this expansion corresponds to the kinetic energy of the material particle in classical mechanics.

*Equations of Motion of Material Particles.* From (43) we obtain, by differentiating by the time $l$, and using the principle of momentum, in the notation of three-dimensional vectors,

$$\mathbf{K} = \frac{d}{dl}\left(\frac{m\mathbf{q}}{\sqrt{1 - q^2}}\right) \tag{46}$$

This equation, which was previously employed by H. A. Lorentz for the motion of electrons, has been proved to be true, with great accuracy, by experiments with $\beta$-rays.

*Energy Tensor of the Electromagnetic Field.* Before the development of the theory of relativity it was known that the principles of energy and momentum could be expressed in a differential form for the electromagnetic field. The four-dimensional formulation of these principles leads to an important conception, that of the energy tensor, which is important for the further development of the theory of relativity.

* The emission of energy in radioactive processes is evidently connected with the fact that the atomic weights are not integers. The equivalence between mass at rest and energy at rest which is expressed in equation (44) has been confirmed in many cases during recent years. In radio-active decomposition the sum of the resulting masses is always less than the mass of the decomposing atom. The difference appears in the form of kinetic energy of the generated particles as well as in the form of released radiational energy.

If in the expression for the 4-vector of force per unit volume,

$$K_\mu = \phi_{\mu\nu} \mathfrak{J}_\nu$$

using the field equations (32), we express $\mathfrak{J}_\mu$ in terms of the field intensities, $\phi_{\mu\nu}$, we obtain, after some transformations and repeated application of the field equations (32) and (33), the expression

$$K_\mu = -\frac{\partial T_{\mu\nu}}{\partial x_\nu} \tag{47}$$

where we have written*

$$T_{\mu\nu} = -\tfrac{1}{4}\phi_{\alpha\beta}^2 \delta_{\mu\nu} + \phi_{\mu\alpha}\phi_{\nu\alpha} \tag{48}$$

The physical meaning of equation (47) becomes evident if in place of this equation we write, using a new notation,

$$\left\{\begin{array}{l} k_x = -\dfrac{\partial p_{xx}}{\partial x} - \dfrac{\partial p_{xy}}{\partial y} - \dfrac{\partial p_{xz}}{\partial z} - \dfrac{\partial (ib_x)}{\partial (il)} \\ \cdot \quad \cdot \quad \cdot \quad \cdot \quad \cdot \quad \cdot \quad \cdot \quad \cdot \\ \cdot \quad \cdot \quad \cdot \quad \cdot \quad \cdot \quad \cdot \quad \cdot \quad \cdot \\ i\lambda = -\dfrac{\partial (is_x)}{\partial x} - \dfrac{\partial (is_y)}{\partial y} - \dfrac{\partial (is_z)}{\partial z} - \dfrac{\partial (-\eta)}{\partial (il)} \end{array}\right. \tag{47a}$$

or, on eliminating the imaginary,

$$\left\{\begin{array}{l} k_x = -\dfrac{\partial p_{xx}}{\partial x} - \dfrac{\partial p_{xy}}{\partial y} - \dfrac{\partial p_{xz}}{\partial z} - \dfrac{\partial b_x}{\partial l} \\ \cdot \quad \cdot \quad \cdot \quad \cdot \quad \cdot \quad \cdot \\ \cdot \quad \cdot \quad \cdot \quad \cdot \quad \cdot \quad \cdot \\ \lambda = -\dfrac{\partial s_x}{\partial x} - \dfrac{\partial s_y}{\partial y} - \dfrac{\partial s_z}{\partial z} - \dfrac{\partial \eta}{\partial l} \end{array}\right. \tag{47b}$$

When expressed in the latter form, we see that the first three equations state the principle of momentum; $p_{xx} \ldots p_{zz}$ are the Maxwell stresses in the electro-magnetic

* To be summed for the indices $\alpha$ and $\beta$.

field, and $(b_x, b_y, b_z)$ is the vector momentum per unit volume of the field. The last of equations (47b) expresses the energy principle; $\mathbf{s}$ is the vector flow of energy, and $\eta$ the energy per unit volume of the field. In fact, we get from (48) by introducing the real components of the field intensity the following expressions well known from electrodynamics:

$$
(48a)\quad \begin{cases}
p_{xx} = -h_x h_x + \frac{1}{2}(h_x^2 + h_y^2 + h_z^2) \\
\qquad\quad -e_x e_x + \frac{1}{2}(e_x^2 + e_y^2 + e_z^2) \\
\qquad\qquad\qquad p_{xy} = -h_x h_y - e_x e_y \\
\qquad\qquad\qquad p_{xz} = -h_x h_z - e_x e_z \\
\cdot\quad\cdot\quad\cdot\quad\cdot\quad\cdot\quad\cdot\quad\cdot\quad\cdot\quad\cdot \\
\cdot\quad\cdot\quad\cdot\quad\cdot\quad\cdot\quad\cdot\quad\cdot\quad\cdot\quad\cdot \\
b_x = s_x = e_y h_z - e_z h_y \\
\cdot\quad\cdot\quad\cdot\quad\cdot\quad\cdot\quad\cdot\quad\cdot\quad\cdot\quad\cdot \\
\cdot\quad\cdot\quad\cdot\quad\cdot\quad\cdot\quad\cdot\quad\cdot\quad\cdot\quad\cdot \\
\eta = +\frac{1}{2}(e_x^2 + e_y^2 + e_z^2 + h_x^2 + h_y^2 + h_z^2)
\end{cases}
$$

We notice from (48) that the energy tensor of the electromagnetic field is symmetrical; with this is connected the fact that the momentum per unit volume and the flow of energy are equal to each other (relation between energy and inertia).

We therefore conclude from these considerations that the energy per unit volume has the character of a tensor. This has been proved directly only for an electromagnetic field, although we may claim universal validity for it. Maxwell's equations determine the electromagnetic field when the distribution of electric charges and currents is known. But we do not know the laws which govern the currents and charges. We do know, indeed, that electricity consists of elementary particles (electrons, positive nuclei), but from a theoretical point of view we cannot comprehend this. We do not know the energy factors which determine the distribution of electricity in particles of definite size and charge, and all attempts to complete the theory in this direction have failed. If then we can build upon Maxwell's equations at all, the energy tensor of the electromagnetic field is known only outside the charged particles.* In these regions, outside of charged particles, the only regions in which we can believe that we have the complete expression for the energy tensor, we have, by (47),

$$\frac{\partial T_{\mu\nu}}{\partial x_\nu} = 0. \tag{47c}$$

*General Expressions for the Conservation Principles.* We can hardly avoid making the assumption that in all other cases, also, the space distribution of energy is given by a symmetrical tensor, $T_{\mu\nu}$, and that this complete energy tensor everywhere satisfies the relation (47c). At any rate we shall see that by means of this assumption we obtain the correct expression for the integral energy principle.

Let us consider a spatially bounded, closed system, which, four-dimensionally, we may represent as a strip, outside of which the $T_{\mu\nu}$ vanish. Integrate equation (47c) over a space section. Since the integrals of $\frac{\partial T_{\mu 1}}{\partial x_1}$, $\frac{\partial T_{\mu 2}}{\partial x_2}$ and $\frac{\partial T_{\mu 3}}{\partial x_3}$ vanish because the $T_{\mu\nu}$ vanish at the limits of integration, we obtain

$$\frac{\partial}{\partial l}\left\{\int T_{\mu 4}dx_1 dx_2 dx_3\right\} = 0 \tag{49}$$

Inside the parentheses are the expressions for the momen-

* It has been attempted to remedy this lack of knowledge by considering the charged particles as proper singularities. But in my opinion this means giving up a real understanding of the structure of matter. It seems to me much better to admit our present inability rather than to be satisfied by a solution that is only apparent.

tum of the whole system, multiplied by $i$, together with the negative energy of the system, so that (49) expresses the conservation principles in their integral form. That this gives the right conception of energy and the conservation principles will be seen from the following considerations.

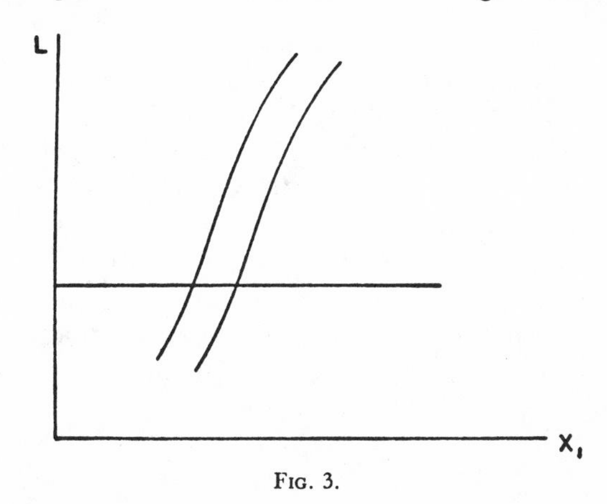


FIG. 3.

## PHENOMENOLOGICAL REPRESENTATION OF THE ENERGY TENSOR OF MATTER

*Hydrodynamical Equations.* We know that matter is built up of electrically charged particles, but we do not know the laws which govern the constitution of these particles. In treating mechanical problems, we are therefore obliged to make use of an inexact description of matter, which corresponds to that of classical mechanics. The density $\sigma$, of a material substance and the hydrodynamical pressures are the fundamental concepts upon which such a description is based.

Let $\sigma_0$ be the density of matter at a place, estimated with reference to a system of co-ordinates moving with the matter. Then $\sigma_0$, the density at rest, is an invariant.

If we think of the matter in arbitrary motion and neglect the pressures (particles of dust *in vacuo*, neglecting the size of the particles and the temperature), then the energy tensor will depend only upon the velocity components, $u_\nu$ and $\sigma_0$. We secure the tensor character of $T_{\mu\nu}$ by putting

$$T_{\mu\nu} = \sigma_0 u_\mu u_\nu \tag{50}$$

in which the $u_\mu$, in the three-dimensional representation, are given by (41). In fact, it follows from (50) that for $q = 0$, $T_{44} = -\sigma_0$ (equal to the negative energy per unit volume), as it should, according to the principle of the equivalence of mass and energy, and according to the physical interpretation of the energy tensor given above. If an external force (four-dimensional vector, $K_\mu$) acts upon the matter, by the principles of momentum and energy the equation

$$K_\mu = \frac{\partial T_{\mu\nu}}{\partial x_\nu}$$

must hold. We shall now show that this equation leads to the same law of motion of a material particle as that already obtained. Let us imagine the matter to be of infinitely small extent in space, that is, a four-dimensional thread; then by integration over the whole thread with respect to the space co-ordinates $x_1$, $x_2$, $x_3$, we obtain

$$\int K_1 dx_1 dx_2 dx_3 = \int \frac{\partial T_{14}}{\partial x_4} dx_1 dx_2 dx_3 = \\ -i \frac{d}{dl} \left\{ \int \sigma_0 \frac{dx_1}{d\tau} \frac{dx_4}{d\tau} dx_1 dx_2 dx_3 \right\}.$$

Now $\int dx_1 dx_2 dx_3 dx_4$ is an invariant, as is, therefore, also $\int \sigma_0 dx_1 dx_2 dx_3 dx_4$. We shall calculate this integral, first with respect to the inertial system which we have chosen, and second, with respect to a system relatively to which the matter has the velocity zero. The integration is to be extended over a filament of the thread for which $\sigma_0$ may be regarded as constant over the whole section. If the space volumes of the filament referred to the two systems are $dV$ and $dV_0$ respectively, then we have

$$\int \sigma_0 dV dl = \int \sigma_0 dV_0 d\tau$$

and therefore also

$$\int \sigma_0 dV = \int \sigma_0 dV_0 \frac{d\tau}{dl} = \int dm\, i\frac{d\tau}{dx_4}.$$

If we substitute the right-hand side for the left-hand side in the former integral, and put $\frac{dx_1}{d\tau}$ outside the sign of integration, we obtain,

$$K_x = \frac{d}{dl}\left(m\frac{dx_1}{d\tau}\right) = \frac{d}{dl}\left(\frac{mq_x}{\sqrt{1-q^2}}\right).$$

We see, therefore, that the generalized conception of the energy tensor is in agreement with our former result.

*The Eulerian Equations for Perfect Fluids.* In order to get nearer to the behaviour of real matter we must add to the energy tensor a term which corresponds to the pressures. The simplest case is that of a perfect fluid in which the pressure is determined by a scalar $p$. Since the tangential stresses $p_{xy}$, etc., vanish in this case, the contribution to the energy tensor must be of the form $p\delta_{\nu\mu}$. We must therefore put

$$T_{\mu\nu} = \sigma u_\mu u_\nu + p\delta_{\mu\nu} \tag{51}$$

At rest, the density of the matter, or the energy per unit volume, is in this case, not $\sigma$ but $\sigma - p$. For

$$-T_{44} = -\sigma \frac{dx_4}{d\tau}\frac{dx_4}{d\tau} - p\delta_{44} = \sigma - p.$$

In the absence of any force, we have

$$\frac{\partial T_{\mu\nu}}{\partial x_\nu} = \sigma u_\nu \frac{\partial u_\mu}{\partial x_\nu} + u_\mu \frac{\partial(\sigma u_\nu)}{\partial x_\nu} + \frac{\partial p}{\partial x_\mu} = 0.$$

If we multiply this equation by $u_\mu \left(= \frac{dx_\mu}{d\tau}\right)$ and sum for the $\mu$'s we obtain, using (40).

$$-\frac{\partial(\sigma u_\nu)}{\partial x_\nu} + \frac{dp}{d\tau} = 0 \tag{52}$$

where we have put $\frac{\partial p}{\partial x_\mu}\frac{dx_\mu}{d\tau} = \frac{dp}{d\tau}$. This is the equation of continuity, which differs from that of classical mechanics by the term $\frac{dp}{d\tau}$, which, practically, is vanishingly small. Observing (52), the conservation principles take the form

$$\sigma \frac{du_\mu}{d\tau} + u_\mu \frac{dp}{d\tau} + \frac{\partial p}{\partial x_\mu} = 0. \tag{53}$$

The equations for the first three indices evidently correspond to the Eulerian equations. That the equations (52) and (53) correspond, to a first approximation, to the hydrodynamical equations of classical mechanics, is a further confirmation of the generalized energy principle. The density of matter (or of energy) has tensor character (specifically, it constitutes a symmetrical tensor).

# THE GENERAL THEORY OF RELATIVITY

ALL of the previous considerations have been based upon the assumption that all inertial systems are equivalent for the description of physical phenomena, but that they are preferred, for the formulation of the laws of nature, to spaces of reference in a different state of motion. We can think of no cause for this preference for definite states of motion to all others, according to our previous considerations, either in the perceptible bodies or in the concept of motion; on the contrary, it must be regarded as an independent property of the space-time continuum. The principle of inertia, in particular, seems to compel us to ascribe physically objective properties to the space-time continuum. Just as it was consistent from the Newtonian standpoint to make both the statements, *tempus est absolutum*, *spatium est absolutum*, so from the standpoint of the special theory of relativity we must say, *continuum spatii et temporis est absolutum*. In this latter statement *absolutum* means not only "physically real," but also "independent in its physical properties, having a physical effect, but not itself influenced by physical conditions."

As long as the principle of inertia is regarded as the keystone of physics, this standpoint is certainly the only one which is justified. But there are two serious criticisms of the ordinary conception. In the first place, it is contrary to the mode of thinking in science to conceive of a thing (the space-time continuum) which acts itself, but which cannot be acted upon. This is the reason why E. Mach was led to make the attempt to eliminate space as an active cause in the system of mechanics. According to him, a material particle does not move in unaccelerated motion relatively to space, but relatively to the centre of all the other masses in the universe; in this way the series of causes of mechanical phenomena was closed, in contrast to the mechanics of Newton and Galileo. In order to develop this idea within the limits of the modern theory of action through a medium, the properties of the space-time continuum which determine inertia must be regarded as field properties of space, analogous to the electromagnetic field. The concepts of classical mechanics afford no way of expressing this. For this reason Mach's attempt at a solution failed for the time being. We shall come back to this point of view later. In the second place, classical mechanics exhibits a deficiency which directly calls for an extension of the principle of relativity to spaces of reference which are not in uniform motion relatively to each other. The ratio of the masses of two bodies is defined in mechanics in two ways which differ from each other fundamentally; in the first place, as the reciprocal ratio of the accelerations which the same motive force imparts to them (inert mass), and in the second place, as the ratio of the forces which act upon them in the same gravitational field (gravitational mass). The equality of these two masses, so differently defined, is a fact which is confirmed by experiments of very high accuracy (experiments of Eötvös), and classical mechanics offers no explanation for this equality. It is, however, clear that science is fully justified in assigning such a numerical equality only after this numerical equality is reduced to an equality of the real nature of the two concepts.

That this object may actually be attained by an extension of the principle of relativity, follows from the following consideration. A little reflection will show that the law of the equality of the inert and the gravitational mass is equivalent to the assertion that the acceleration imparted to a body by a gravitational field is independent of the nature of the body. For Newton's equation of motion in a gravitational field, written out in full, is

(Inert mass) . (Acceleration) = (Intensity of the gravitational field) . (Gravitational mass).

It is only when there is numerical equality between the inert and gravitational mass that the acceleration is independent of the nature of the body. Let now $K$ be an inertial system. Masses which are sufficiently far from each other and from other bodies are then, with respect to $K$, free from acceleration. We shall also refer these masses to a system of co-ordinates $K'$, uniformly accelerated with respect to $K$. Relatively to $K'$ all the masses have equal and parallel accelerations; with respect to $K'$ they behave just as if a gravitational field were present and $K'$ were unaccelerated. Overlooking for the present the question as to the "cause" of such a gravitational field, which will occupy us later, there is nothing to prevent our conceiving this gravitational field as real, that is, the conception that $K'$ is "at rest" and a gravitational field is present we may consider as equivalent to the conception that only $K$ is an "allowable" system of co-ordinates and no gravitational field is present. The assumption of the complete physical equivalence of the systems of coordinates, $K$ and $K'$, we call the "principle of equivalence;"

this principle is evidently intimately connected with the law of the equality between the inert and the gravitational mass, and signifies an extension of the principle of relativity to co-ordinate systems which are in non-uniform motion relatively to each other. In fact, through this conception we arrive at the unity of the nature of inertia and gravitation. For according to our way of looking at it, the same masses may appear to be either under the action of inertia alone (with respect to $K$) or under the combined action of inertia and gravitation (with respect to $K'$). The possibility of explaining the numerical equality of inertia and gravitation by the unity of their nature gives to the general theory of relativity, according to my conviction, such a superiority over the conceptions of classical mechanics, that all the difficulties encountered must be considered as small in comparison with this progress.

What justifies us in dispensing with the preference for inertial systems over all other co-ordinate systems, a preference that seems so securely established by experience? The weakness of the principle of inertia lies in this, that it involves an argument in a circle: a mass moves without acceleration if it is sufficiently far from other bodies; we know that it is sufficiently far from other bodies only by the fact that it moves without acceleration. Are there at all any inertial systems for very extended portions of the space-time continuum, or, indeed, for the whole universe? We may look upon the principle of inertia as established, to a high degree of approximation, for the space of our planetary system, provided that we neglect the perturbations due to the sun and planets. Stated more exactly, there are finite regions, where, with respect to a suitably chosen space of reference, material particles move freely without acceleration, and in which the laws of the special theory of relativity, which have been developed above, hold with remarkable accuracy. Such regions we shall call "Galilean regions." We shall proceed from the consideration of such regions as a special case of known properties.

The principle of equivalence demands that in dealing with Galilean regions we may equally well make use of non-inertial systems, that is, such co-ordinate systems as, relatively to inertial systems, are not free from acceleration and rotation. If, further, we are going to do away completely with the vexing question as to the objective reason for the preference of certain systems of co-ordinates, then we must allow the use of arbitrarily moving systems of co-ordinates. As soon as we make this attempt seriously we come into conflict with that physical interpretation of space and time to which we were led by the special theory of relativity. For let $K'$ be a system of co-ordinates whose $z'$-axis coincides with the $z$-axis of $K$, and which rotates about the latter axis with constant angular velocity. Are the configurations of rigid bodies, at rest relatively to $K'$, in accordance with the laws of Euclidean geometry? Since $K'$ is not an inertial system, we do not know directly the laws of configuration of rigid bodies with respect to $K'$, nor the laws of nature, in general. But we do know these laws with respect to the inertial system $K$, and we can therefore infer their form with respect to $K'$. Imagine a circle drawn about the origin in the $x'y'$ plane of $K'$, and a diameter of this circle. Imagine, further, that we have given a large number of rigid rods, all equal to each other. We suppose these laid in series along the periphery and the diameter of the circle, at rest relatively to $K'$. If $U$ is the number of these rods along the periphery, $D$ the number along the diameter, then, if $K'$ does not rotate relatively to $K$, we shall have

$$\frac{U}{D} = \pi.$$

But if $K'$ rotates we get a different result. Suppose that at a definite time $t$, of $K$ we determine the ends of all the rods. With respect to $K$ all the rods upon the periphery experience the Lorentz contraction, but the rods upon the diameter do not experience this contraction (along their lengths!).* It therefore follows that

$$\frac{U}{D} > \pi.$$

It therefore follows that the laws of configuration of rigid bodies with respect to $K'$ do not agree with the laws of configuration of rigid bodies that are in accordance with Euclidean geometry. If, further, we place two similar clocks (rotating with $K'$), one upon the periphery, and the other at the centre of the circle, then, judged from $K$, the clock on the periphery will go slower than the clock at the centre. The same thing must take place, judged from $K'$, if we do not define time with respect to $K'$ in a wholly unnatural way, (that is, in such a way that the laws with respect to $K'$ depend explicitly upon the time). Space and time, therefore, cannot be defined with respect to $K'$ as they were in the special theory of relativity with respect to inertial systems. But, according to the principle of equivalence, $K'$ may also be considered as a system at rest, with respect to which there is a gravitational field (field of centrifugal force, and

* These considerations assume that the behavior of rods and clocks depends only upon velocities, and not upon accelerations, or, at least, that the influence of acceleration does not counteract that of velocity.

force of Coriolis). We therefore arrive at the result: the gravitational field influences and even determines the metrical laws of the space-time continuum. If the laws of configuration of ideal rigid bodies are to be expressed geometrically, then in the presence of a gravitational field the geometry is not Euclidean.

The case that we have been considering is analogous to that which is presented in the two-dimensional treatment of surfaces. It is impossible in the latter case also, to introduce co-ordinates on a surface (e.g. the surface of an ellipsoid) which have a simple metrical significance, while on a plane the Cartesian co-ordinates, $x_1$, $x_2$, signify directly lengths measured by a unit measuring rod. Gauss overcame this difficulty, in his theory of surfaces, by introducing curvilinear co-ordinates which, apart from satisfying conditions of continuity, were wholly arbitrary, and only afterwards these co-ordinates were related to the metrical properties of the surface. In an analogous way we shall introduce in the general theory of relativity arbitrary co-ordinates, $x_1$, $x_2$, $x_3$, $x_4$, which shall number uniquely the space-time points, so that neighbouring events are associated with neighbouring values of the co-ordinates; otherwise, the choice of co-ordinates is arbitrary. We shall be true to the principle of relativity in its broadest sense if we give such a form to the laws that they are valid in every such four-dimensional system of co-ordinates, that is, if the equations expressing the laws are co-variant with respect to arbitrary transformations.

The most important point of contact between Gauss's theory of surfaces and the general theory of relativity lies in the metrical properties upon which the concepts of both theories, in the main, are based. In the case of the theory of surfaces, Gauss's argument is as follows.

Plane geometry may be based upon the concept of the distance $ds$, between two infinitely near points. The concept of this distance is physically significant because the distance can be measured directly by means of a rigid measuring rod. By a suitable choice of Cartesian co-ordinates this distance may be expressed by the formula $ds^2 = dx_1^2 + dx_2^2$. We may base upon this quantity the concepts of the straight line as the geodesic ($\delta\int ds = 0$), the interval, the circle, and the angle, upon which the Euclidean plane geometry is built. A geometry may be developed upon another continuously curved surface, if we observe that an infinitesimally small portion of the surface may be regarded as plane, to within relatively infinitesimal quantities. There are Cartesian co-ordinates, $X_1$, $X_2$, upon such a small portion of the surface, and the distance between two points, measured by a measuring rod, is given by

$$ds^2 = dX_1^2 + dX_2^2.$$

If we introduce arbitrary curvilinear co-ordinates, $x_1$, $x_2$, on the surface, then $dX_1$, $dX_2$, may be expressed linearly in terms of $dx_1$, $dx_2$. Then everywhere upon the surface we have

$$ds^2 + g_{11}dx_1^2 + 2g_{12}dx_1dx_2 + g_{22}dx_2^2$$

where $g_{11}$, $g_{12}$, $g_{22}$ are determined by the nature of the surface and the choice of co-ordinates; if these quantities are known, then it is also known how networks of rigid rods may be laid upon the surface. In other words, the geometry of surfaces may be based upon this expression for $ds^2$ exactly as plane geometry is based upon the corresponding expression.

There are analogous relations in the four-dimensional space-time continuum of physics. In the immediate neighbourhood of an observer, falling freely in a gravitational field, there exists no gravitational field. We can therefore always regard an infinitesimally small region of the space-time continuum as Galilean. For such an infinitely small region there will be an inertial system (with the space co-ordinates, $X_1$, $X_2$, $X_3$, and the time co-ordinate $X_4$) relatively to which we are to regard the laws of the special theory of relativity as valid. The quantity which is directly measurable by our unit measuring rods and clocks,

$$dX_1^2 + dX_2^2 + dX_3^2 - dX_4^2$$

or its negative,

$$ds^2 = -dX_1^2 - dX_2^2 - dX_3^2 + dX_4^2 \tag{54}$$

is therefore a uniquely determinate invariant for two neighbouring events (points in the four-dimensional continuum), provided that we use measuring rods that are equal to each other when brought together and superimposed, and clocks whose rates are the same when they are brought together. In this the physical assumption is essential that the relative lengths of two measuring rods and the relative rates of two clocks are independent, in principle, of their previous history. But this assumption is certainly warranted by experience; if it did not hold there could be no sharp spectral lines, since the single atoms of the same element certainly do not have the same history, and since—on the assumption of relative variability of the single atoms depending on previous history—it would be absurd to suppose that the masses or proper frequencies of these atoms ever had been equal to one another.

Space-time regions of finite extent are, in general, not Galilean, so that a gravitational field cannot be done away with by any choice of co-ordinates in a finite region. There is, therefore, no choice of co-ordinates for which the metrical relations of the special theory of relativity hold in a finite region. But the invariant $ds$ always exists for two neighbouring points (events) of the continuum. This invariant $ds$ may be expressed in arbitrary co-ordinates. If one observes that the local $dX_\nu$ may be expressed linearly in terms of the co-ordinate differentials $dx_\nu$, $ds^2$ may be expressed in the form

$$ds^2 = g_{\mu\nu} dx_\mu dx_\nu. \tag{55}$$

The functions $g_{\mu\nu}$ describe, with respect to the arbitrarily chosen system of co-ordinates, the metrical relations of the space-time continuum and also the gravitational field. As in the special theory of relativity, we have to discriminate between time-like and space-like line elements in the four-dimensional continuum; owing to the change of sign introduced, time-like line elements have a real, space-like line elements an imaginary $ds$. The time-like $ds$ can be measured directly by a suitably chosen clock.

According to what has been said, it is evident that the formulation of the general theory of relativity requires a generalization of the theory of invariants and the theory of tensors; the question is raised as to the form of the equations which are co-variant with respect to arbitrary point transformations. The generalized calculus of tensors was developed by mathematicians long before the theory of relativity. Riemann first extended Gauss's train of thought to continua of any number of dimensions; with prophetic vision he saw the physical meaning of this generalization of Euclid's geometry. Then followed the development of the theory in the form of the calculus of tensors, particularly by Ricci and Levi-Civita. This is the place for a brief presentation of the most important mathematical concepts and operations of this calculus of tensors.

We designate four quantities, which are defined as functions of the $x_\nu$ with respect to every system of coordinates, as components, $A^\nu$, of a contra-variant vector, if they transform in a change of co-ordinates as the co-ordinate differentials $dx_\nu$. We therefore have

$$A^{\mu'} = \frac{\partial x'_\mu}{\partial x_\nu} A^\nu. \tag{56}$$

Besides these contra-variant vectors, there are also co-variant vectors. If $B_\nu$ are the components of a co-variant vector, these vectors are transformed according to the rule

$$B'_\mu = \frac{\partial x_\nu}{\partial x'_\mu} B_\nu. \tag{57}$$

The definition of a co-variant vector is chosen in such a way that a co-variant vector and a contra-variant vector together form a scalar according to the scheme,

$$\phi = B_\nu A^\nu \text{ (summed over the } \nu\text{)}.$$

For we have

$$B'_\mu A^{\mu'} = \frac{\partial x_\alpha}{\partial x'_\mu} \frac{\partial x'_\mu}{\partial x_\beta} B_\alpha A^\beta = B_\alpha A^\alpha.$$

In particular, the derivatives $\frac{\partial \phi}{\partial x_\alpha}$ of a scalar $\phi$, are components of a co-variant vector, which, with the co-ordinate differentials, form the scalar $\frac{\partial \phi}{\partial x_\alpha} dx_\alpha$; we see from this example how natural is the definition of the co-variant vectors.

There are here, also, tensors of any rank, which may have co-variant or contra-variant character with respect to each index; as with vectors, the character is designated by the position of the index. For example, $A_\mu^\nu$ denotes a tensor of the second rank, which is co-variant with respect to the index $\mu$, and contra-variant with respect to the index $\nu$. The tensor character indicates that the equation of transformation is

$$A_\mu^{\nu'} = \frac{\partial x_\alpha}{\partial x'_\mu} \frac{\partial x'_\nu}{\partial x_\beta} A_\alpha^\beta. \tag{58}$$

Tensors may be formed by the addition and subtraction of tensors of equal rank and like character, as in the theory of invariants of orthogonal linear substitutions, for example,

$$A_\mu^\nu + B_\mu^\nu = C_\mu^\nu. \tag{59}$$

The proof of the tensor character of $C_\mu^\nu$ depends upon (58).

Tensors may be formed by multiplication, keeping the character of the indices, just as in the theory of invariants of linear orthogonal transformations, for example,

$$A_\mu^\nu B_{\sigma\tau} = C_{\mu\sigma\tau}^\nu. \tag{60}$$

The proof follows directly from the rule of transformation.

Tensors may be formed by contraction with respect to two indices of different character, for example,

$$A_{\mu\sigma\tau}^\mu = B_{\sigma\tau}. \tag{61}$$

The tensor character of $A_{\mu\sigma\tau}^\mu$ determines the tensor character of $B_{\sigma\tau}$. Proof—

$$A_{\mu\sigma\tau}^{\mu'} = \frac{\partial x_\alpha}{\partial x'_\mu} \frac{\partial x'_\mu}{\partial x_\beta} \frac{\partial x_s}{\partial x'_\sigma} \frac{\partial x_t}{\partial x'_\tau} A_{\alpha st}^\beta = \frac{\partial x_s}{\partial x'_\sigma} \frac{\partial x_t}{\partial x'_\tau} A_{\alpha st}^\alpha.$$

The properties of symmetry and skew-symmetry of a tensor with respect to two indices of like character have the same significance as in the theory of special relativity.

With this, everything essential has been said with regard to the algebraic properties of tensors.

*The Fundamental Tensor.* It follows from the invariance of $ds^2$ for an arbitrary choice of the $dx_\nu$, in connexion with the condition of symmetry consistent with (55), that the $g_{\mu\nu}$ are components of a symmetrical co-variant tensor (Fundamental Tensor). Let us form the determinant, $g$, of the $g_{\mu\nu}$, and also the cofactors, divided by $g$, corresponding to the various $g_{\mu\nu}$. These cofactors, divided by $g$, will be denoted by $g^{\mu\nu}$, and their co-variant character is not yet known. Then we have

$$g_{\mu\alpha}g^{\mu\beta} = \delta_\alpha^\beta = \begin{matrix} 1 \text{ if } \alpha = \beta \\ 0 \text{ if } \alpha \neq \beta \end{matrix} \tag{62}$$

If we form the infinitely small quantities (co-variant vectors)

$$d\xi_\mu = g_{\mu\alpha}dx_\alpha \tag{63}$$

multiply by $g^{\mu\beta}$ and sum over the $\mu$, we obtain, by the use of (62),

$$dx_\beta = g^{\beta\mu}d\xi_\mu. \tag{64}$$

Since the ratios of the $d\xi_\mu$ are arbitrary, and the $dx_\beta$ as well as the $d\xi_\mu$ are components of vectors, it follows that the $g^{\mu\nu}$ are the components of a contra-variant tensor* (contra-variant fundamental tensor). The tensor character of $\delta_\alpha^\beta$ (mixed fundamental tensor) accordingly follows,

* If we multiply (64) by $\frac{\partial x'_\alpha}{\partial x_\beta}$, sum over the $\beta$, and replace the $d\xi_\mu$ by a transformation to the accented system, we obtain

$$dx'_\alpha = \frac{\partial x'_\sigma}{\partial x_\mu}\frac{\partial x'_\alpha}{\partial x_\beta}g^{\mu\beta}d\xi'_\sigma.$$

The statement made above follows from this, since, by (64), we must also have $dx'_\alpha = g^{\sigma\alpha'}d\xi'_\sigma$, and both equations must hold for every choice of the $d\xi'_\sigma$.

by (62). By means of the fundamental tensor, instead of tensors with co-variant index character, we can introduce tensors with contra-variant index character, and conversely. For example,

$$\begin{aligned} A^{\mu} &= g^{\mu\alpha} A_{\alpha} \\ A_{\mu} &= g_{\mu\alpha} A^{\alpha} \\ T^{\sigma}_{\mu} &= g^{\sigma\nu} T_{\mu\nu}. \end{aligned}$$

*Volume Invariants.* The volume element

$$\int dx_1 dx_2 dx_3 dx_4 = dx$$

is not an invariant. For by Jacobi's theorem,

$$dx' = \left| \frac{dx'_{\mu}}{dx_{\nu}} \right| dx. \tag{65}$$

But we can complement $dx$ so that it becomes an invariant. If we form the determinant of the quantities

$$g'_{\mu\nu} = \frac{\partial x_{\alpha}}{\partial x'_{\mu}} \frac{\partial x_{\beta}}{\partial x'_{\nu}} g_{\alpha\beta}$$

we obtain, by a double application of the theorem of multiplication of determinants,

$$g' = |g'_{\mu\nu}| = \left| \frac{\partial x_{\nu}}{\partial x'_{\mu}} \right|^{2} \cdot |g_{\mu\nu}| = \left| \frac{\partial x'_{\mu}}{\partial x_{\nu}} \right|^{-2} g.$$

We therefore get the invariant,

$$\sqrt{g'}\, dx' = \sqrt{g}\, dx.$$

*Formation of Tensors by Differentiation.* Although the algebraic operations of tensor formation have proved to be as simple as in the special case of invariance with respect to linear orthogonal transformations, nevertheless in the general case, the invariant differential operations are, unfortunately, considerably more complicated. The reason for this is as follows. If $A^\mu$ is a contra-variant vector, the coefficients of its transformation, $\frac{\partial x'_\mu}{\partial x_\nu}$, are independent of position only if the transformation is a linear one. Then the vector components, $A^\mu + \frac{\partial A^\mu}{\partial x_\alpha} dx_\alpha$, at a neighbouring point transform in the same way as the $A^\mu$, from which follows the vector character of the vector differentials, and the tensor character of $\frac{\partial A^\mu}{\partial x_\alpha}$. But if the $\frac{\partial x'_\mu}{\partial x_\nu}$ are variable this is no longer true.

That there are, nevertheless, in the general case, invariant differential operations for tensors, is recognized most satisfactorily in the following way, introduced by Levi-Civita and Weyl. Let $(A^\mu)$ be a contra-variant vector whose components are given with respect to the co-ordinate system of the $x_\nu$. Let $P_1$ and $P_2$ be two infinitesimally near points of the continuum. For the infinitesimal region surrounding the point $P_1$, there is, according to our way of considering the matter, a co-ordinate system of the $X_\nu$ (with imaginary $X_4$-co-ordinate ) for which the continuum is Euclidean. Let $A^\mu_{(1)}$ be the co-ordinates of the vector at the point $P_1$. Imagine a vector drawn at the point $P_2$, using the local system of the $X_\nu$, with the same co-ordinates (parallel vector through $P_2$), then this parallel vector is uniquely determined by the vector at $P_1$ and the displacement. We designate this operation, whose uniqueness will appear in the sequel, the parallel displacement of the vector $A_\mu$ from $P_1$ to the infinitesimally near point $P_2$. If we form the vector difference of the vector $(A^\mu)$ at the point $P_2$ and the vector obtained by parallel displacement from

$P_1$ to $P_2$, we get a vector which may be regarded as the differential of the vector $(A^\mu)$ for the given displacement $(dx_\nu)$.

This vector displacement can naturally also be considered with respect to the co-ordinate system of the $x_\nu$. If $A^\nu$ are the co-ordinates of the vector at $P_1$, $A^\nu + \delta A^\nu$ the co-ordinates of the vector displaced to $P_2$ along the interval $(dx_\nu)$, then the $\delta A^\nu$ do not vanish in this case. We know of these quantities, which do not have a vector character, that they must depend linearly and homogeneously upon the $dx_\nu$ and the $A^\nu$. We therefore put

$$\delta A^\nu = -\Gamma^\nu_{\alpha\beta} A^\alpha dx_\beta. \tag{67}$$

In addition, we can state that the $\Gamma^\nu_{\alpha\beta}$ must be symmetrical with respect to the indices $\alpha$ and $\beta$. For we can assume from a representation by the aid of a Euclidean system of local co-ordinates that the same parallelogram will be described by the displacement of an element $d^{(1)}x_\nu$ along a second element $d^{(2)}x_\nu$ as by a displacement of $d^{(2)}x_\nu$ along $d^{(1)}x_\nu$. We must therefore have

$$\begin{aligned} d^{(2)}x_\nu + (d^{(1)}x_\nu - \Gamma^\nu_{\alpha\beta} d^{(1)}x_\alpha d^{(2)}x_\beta) \\ = d^{(1)}x_\nu + (d^{(2)}x_\nu - \Gamma^\nu_{\alpha\beta} d^{(2)}x_\alpha d^{(1)}x_\beta). \end{aligned}$$

The statement made above follows from this, after interchanging the indices of summation, $\alpha$ and $\beta$, on the right-hand side.

Since the quantities $g_{\mu\nu}$ determine all the metrical properties of the continuum, they must also determine the $\Gamma^\nu_{\alpha\beta}$. If we consider the invariant of the vector $A^\nu$, that is, the square of its magnitude,

$$g_{\mu\nu} A^\mu A^\nu$$

which is an invariant, this cannot change in a parallel displacement. We therefore have

$$0 = \delta(g_{\mu\nu}A^\mu A^\nu) = \frac{\partial g_{\mu\nu}}{\partial x_\alpha} A^\mu A^\nu dx_\alpha + g_{\mu\nu}A^\mu \delta A^\nu + g_{\mu\nu}A^\nu \delta A^\mu$$

or, by (67),

$$\left(\frac{\partial g_{\mu\nu}}{\partial x_\alpha} - g_{\mu\beta}\Gamma^\beta_{\nu\alpha} - g_{\nu\beta}\Gamma^\beta_{\mu\alpha}\right) A^\mu A^\nu dx_\alpha = 0.$$

Owing to the symmetry of the expression in the brackets with respect to the indices $\mu$ and $\nu$, this equation can be valid for an arbitrary choice of the vectors $(A^\mu)$ and $dx_\nu$ only when the expression in the brackets vanishes for all combinations of the indices. By a cyclic interchange of the indices $\mu$, $\nu$, $a$, we obtain thus altogether three equations, from which we obtain, on taking into account the symmetrical property of the $\Gamma^\alpha_{\mu\nu}$,

$$\left[{\mu\nu \atop \alpha}\right] = g_{\alpha\beta}\Gamma^\beta_{\mu\nu} \tag{68}$$

in which, following Christoffel, the abbreviation has been used,

$$\left[{\mu\nu \atop \alpha}\right] = \tfrac{1}{2}\left(\frac{\partial g_{\mu\alpha}}{\partial x_\nu} + \frac{\partial g_{\nu\alpha}}{\partial x_\mu} - \frac{\partial g_{\mu\nu}}{\partial x_\alpha}\right). \tag{69}$$

If we multiply (68) by $g^{\alpha\sigma}$ and sum over the $\alpha$, we obtain

$$\Gamma^\sigma_{\mu\nu} = \tfrac{1}{2}g^{\sigma\alpha}\left(\frac{\partial g_{\mu\alpha}}{\partial x_\nu} + \frac{\partial g_{\nu\alpha}}{\partial x_\mu} - \frac{\partial g_{\mu\nu}}{\partial x_\alpha}\right) = \left\{{\mu\nu \atop \sigma}\right\} \tag{70}$$

in which $\left\{{\mu\nu \atop \sigma}\right\}$ is the Christoffel symbol of the second kind. Thus the quantities $\Gamma$ are deduced from the $g_{\mu\nu}$. Equations (67) and (70) are the foundation for the following discussion.

*Co-variant Differentiation of Tensors.* If $(A^\mu + \delta A^\mu)$ is the vector resulting from an infinitesimal parallel displacement from $P_1$ to $P_2$, and $(A^\mu + dA^\mu)$ the vector $A^\mu$ at the point $P_2$, then the difference of these two,

$$dA^\mu - \delta A^\mu = \left(\frac{\partial A^\mu}{\partial x_\sigma} + \Gamma^\mu_{\sigma\alpha} A^\alpha\right) dx_\sigma$$

is also a vector. Since this is the case for an arbitrary choice of the $dx_\sigma$, it follows that

$$A^\mu_{;\sigma} = \frac{\partial A^\mu}{\partial x_\sigma} + \Gamma^\mu_{\sigma\alpha} A^\alpha \tag{71}$$

is a tensor, which we designate as the co-variant derivative of the tensor of the first rank (vector). Contracting this tensor, we obtain the divergence of the contra-variant tensor $A^\mu$. In this we must observe that according to (70),

$$\Gamma^\sigma_{\mu\sigma} = \tfrac{1}{2} g^{\sigma\alpha} \frac{\partial g_{\sigma\alpha}}{\partial x_\mu} = \frac{1}{\sqrt{g}} \frac{\partial \sqrt{g}}{\partial x_\mu} \tag{72}$$

If we put, further,

$$A^\mu \sqrt{g} = \mathfrak{A}^\mu \tag{73}$$

a quantity designated by Weyl as the contra-variant tensor density* of the first rank, it follows that,

$$\mathfrak{A} = \frac{\partial \mathfrak{A}^\mu}{\partial x_\mu} \tag{74}$$

is a scalar density.

We get the law of parallel displacement for the co-variant vector $B_\mu$ by stipulating that the parallel displacement

* This expression is justified, in that $A^\mu \sqrt{g}\, dx = \mathfrak{A}^\mu dx$ has a tensor character. Every tensor, when multiplied by $\sqrt{g}$, changes into a tensor density. We employ capital Gothic letters for tensor densities.

shall be effected in such a way that the scalar

$$\phi = A^{\mu}B_{\mu}$$

remains unchanged, and that therefore

$$A^{\mu}\delta B_{\mu} + B_{\mu}\delta A^{\mu}$$

vanishes for every value assigned to $(A^{\mu})$. We therefore get

$$\delta B_{\mu} = \Gamma^{\alpha}_{\mu\sigma} A_{\alpha} dx_{\sigma}. \tag{75}$$

From this we arrive at the co-variant derivative of the co-variant vector by the same process as that which led to (71),

$$B_{\mu\,;\,\sigma} = \frac{\partial B_{\mu}}{\partial x_{\sigma}} - \Gamma^{\alpha}_{\mu\sigma} B_{\alpha}. \tag{76}$$

By interchanging the indices $\mu$ and $\sigma$, and subtracting, we get the skew-symmetrical tensor,

$$\phi_{\mu\sigma} = \frac{\partial B_{\mu}}{\partial x_{\sigma}} - \frac{\partial B_{\sigma}}{\partial x_{\mu}}. \tag{77}$$

For the co-variant differentiation of tensors of the second and higher ranks we may use the process by which (75) was deduced. Let, for example, $(A_{\sigma\tau})$ be a co-variant tensor of the second rank. Then $A_{\sigma\tau}E^{\sigma}F^{\tau}$ is a scalar, if $E$ and $F$ are vectors. This expression must not be changed by the $\delta$-displacement; expressing this by a formula, we get, using (67), $\delta A_{\sigma\tau}$, whence we get the desired co-variant derivative,

$$A_{\sigma\tau;\rho} = \frac{\partial A_{\sigma\tau}}{\partial x_{\rho}} - \Gamma^{\alpha}_{\sigma\rho} A_{\alpha\tau} - \Gamma^{\alpha}_{\tau\rho} A_{\sigma\alpha}. \tag{78}$$

In order that the general law of co-variant differentiation of tensors may be clearly seen, we shall write down two co-variant derivatives deduced in an analogous way:

$$A^{\tau}_{\sigma;\rho} = \frac{\partial A^{\tau}_{\sigma}}{\partial x_{\rho}} - \Gamma^{\alpha}_{\sigma\rho} A^{\tau}_{\alpha} + \Gamma^{\tau}_{\alpha\rho} A^{\alpha}_{\sigma}. \tag{79}$$

$$A^{\sigma\tau}_{;\rho} = \frac{\partial A^{\sigma\tau}}{\partial x_{\rho}} + \Gamma^{\sigma}_{\alpha\rho} A^{\alpha\tau} + \Gamma^{\tau}_{\alpha\rho} A^{\sigma\alpha} \tag{80}$$

The general law of formation now becomes evident. From these formulæ we shall deduce some others which are of interest for the physical applications of the theory.

In case $A_{\sigma\tau}$ is skew-symmetrical, we obtain the tensor

$$A_{\sigma\tau\rho} = \frac{\partial A_{\sigma\tau}}{\partial x_{\rho}} + \frac{\partial A_{\tau\rho}}{\partial x_{\sigma}} + \frac{\partial A_{\rho\sigma}}{\partial x_{\tau}} \tag{81}$$

which is skew-symmetrical in all pairs of indices, by cyclic interchanges and addition.

If, in (78), we replace $A_{\sigma\tau}$ by the fundamental tensor, $g_{\sigma\tau}$, then the right-hand side vanishes identically; an analogous statement holds for (80) with respect to $g^{\sigma\tau}$; that is, the co-variant derivatives of the fundamental tensor vanish. That this must be so we see directly in the local system of co-ordinates.

In case $A^{\sigma\tau}$ is skew-symmetrical, we obtain from (80), by contraction with respect to $\tau$ and $\rho$,

$$\mathfrak{A}^{\sigma} = \frac{\partial \mathfrak{A}^{\sigma\tau}}{\partial x_{\tau}}. \tag{82}$$

In the general case, from (79) and (80), by contraction with respect to $\tau$ and $\rho$, we obtain the equations,

$$\mathfrak{A}_{\sigma} = \frac{\partial \mathfrak{A}^{\alpha}_{\sigma}}{\partial x_{\alpha}} - \Gamma^{\alpha}_{\sigma\beta} \mathfrak{A}^{\beta}_{\alpha}. \tag{83}$$

$$\mathfrak{A}^{\sigma} = \frac{\partial \mathfrak{A}^{\sigma\alpha}}{\partial x_{\alpha}} + \Gamma^{\sigma}_{\alpha\beta} \mathfrak{A}^{\alpha\beta}. \tag{84}$$

*The Riemann Tensor.* If we have given a curve extending from the point $P$ to the point $G$ of the continuum, then a vector $A^\mu$, given at $P$, may, by a parallel displacement, be moved along the curve to $G$. If the continuum is Euclidean (more generally, if by a suitable choice of co-ordinates the $g_{\mu\nu}$ are constants) then the vector obtained at $G$ as a result of this displacement does not depend upon the choice of the curve joining $P$ and $G$. But otherwise, the result depends upon the path of the displacement. In this case,

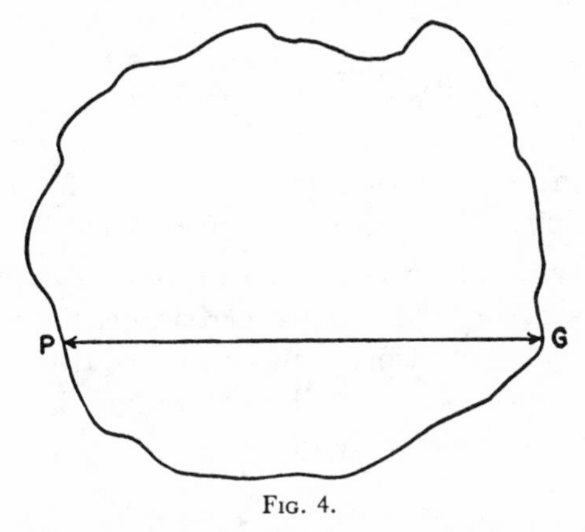


FIG. 4.

therefore, a vector suffers a change, $\Delta A^\mu$ (in its direction, not its magnitude), when it is carried from a point $P$ of a closed curve, along the curve, and back to $P$. We shall now calculate this vector change:

$$\Delta A^\mu = \oint \delta A^\mu.$$

As in Stokes' theorem for the line integral of a vector around a closed curve, this problem may be reduced to the integration around a closed curve with infinitely small linear dimensions; we shall limit ourselves to this case.

We have, first, by (67),

$$\Delta A^{\mu} = -\oint \Gamma^{\mu}_{\alpha\beta} A^{\alpha} dx_{\beta}.$$

In this, $\Gamma^{\mu}_{\alpha\beta}$ is the value of this quantity at the variable point $G$ of the path of integration. If we put

$$\xi^{\mu} = (x_{\mu})_G - (x_{\mu})_P$$

and denote the value of $\Gamma^{\mu}_{\alpha\beta}$ at $P$ by $\overline{\Gamma^{\mu}_{\alpha\beta}}$, then we have, with sufficient accuracy,

$$\Gamma^{\mu}_{\alpha\beta} = \overline{\Gamma^{\mu}_{\alpha\beta}} + \frac{\overline{\partial \Gamma^{\mu}_{\alpha\beta}}}{\partial x_{\nu}} \xi^{\nu}.$$

Let, further, $A^{\alpha}$ be the value obtained from $\overline{A^{\alpha}}$ by a parallel displacement along the curve from $P$ to $G$. It may now easily be proved by means of (67) that $A^{\mu} - \overline{A^{\mu}}$ is infinitely small of the first order, while, for a curve of infinitely small dimensions of the first order, $\Delta A^{\mu}$ is infinitely small of the second order. Therefore there is an error of only the second order if we put

$$A^{\alpha} = \overline{A^{\alpha}} - \overline{\Gamma^{\alpha}_{\sigma\tau}}\,\overline{A^{\sigma}}\,\overline{\xi^{\tau}}.$$

If we introduce these values of $\Gamma^{\mu}_{\alpha\beta}$ and $A^{\alpha}$ into the integral, we obtain, neglecting all quantities of a higher order than the second,

$$\Delta A^{\mu} = -\left(\frac{\partial \Gamma^{\mu}_{\sigma\beta}}{\partial x_{\alpha}} - \Gamma^{\mu}_{\rho\beta}\Gamma^{\rho}_{\sigma\alpha}\right) A^{\sigma} \oint \xi^{\alpha} d\xi^{\beta}. \qquad (85)$$

The quantity removed from under the sign of integration refers to the point $P$. Subtracting $\frac{1}{2}d(\xi^{\alpha}\xi^{\beta})$ from the integrand, we obtain

$$\tfrac{1}{2}\oint (\xi^{\alpha} d\xi^{\beta} - \xi^{\beta} d\xi^{\alpha}).$$

This skew-symmetrical tensor of the second rank, $f^{\alpha\beta}$, characterizes the surface element bounded by the curve in magnitude and position. If the expression in the brackets in (85) were skew-symmetrical with respect to the indices $\alpha$ and $\beta$, we could conclude its tensor character from (85). We can accomplish this by interchanging the summation indices $\alpha$ and $\beta$ in (85) and adding the resulting equation to (85). We obtain

$$2\Delta A^{\mu} = -R^{\mu}_{\sigma\alpha\beta} A^{\sigma} f^{\alpha\beta} \tag{86}$$

in which

$$R^{\mu}_{\sigma\alpha\beta} = -\frac{\partial \Gamma^{\mu}_{\sigma\alpha}}{\partial x_{\beta}} + \frac{\partial \Gamma^{\mu}_{\sigma\beta}}{\partial x_{\alpha}} + \Gamma^{\mu}_{\rho\alpha}\Gamma^{\rho}_{\sigma\beta} - \Gamma^{\mu}_{\rho\beta}\Gamma^{\rho}_{\sigma\alpha}. \tag{87}$$

The tensor character of $R^{\mu}_{\sigma\alpha\beta}$ follows from (86); this is the Riemann curvature tensor of the fourth rank, whose properties of symmetry we do not need to go into. Its vanishing is a sufficient condition (disregarding the reality of the chosen co-ordinates) that the continuum is Euclidean.

By contraction of the Riemann tensor with respect to the indices $\mu$, $\beta$, we obtain the symmetrical tensor of the second rank,

$$R_{\mu\nu} = -\frac{\partial \Gamma^{\alpha}_{\mu\nu}}{\partial x_{\alpha}} + \Gamma^{\alpha}_{\mu\beta}\Gamma^{\beta}_{\nu\alpha} + \frac{\partial \Gamma^{\alpha}_{\mu\alpha}}{\partial x_{\nu}} - \Gamma^{\alpha}_{\mu\nu}\Gamma^{\beta}_{\alpha\beta}. \tag{88}$$

The last two terms vanish if the system of co-ordinates is so chosen that $g$ = constant. From $R_{\mu\nu}$ we can form the scalar,

$$R = g^{\mu\nu} R_{\mu\nu}. \tag{89}$$

*Straightest (Geodesic) Lines.* A line may be constructed in such a way that its successive elements arise from each other by parallel displacements. This is the natural generalization of the straight line of the Euclidean geometry. For such a line, we have

$$\delta\left(\frac{dx_\mu}{ds}\right) = -\Gamma^\mu_{\alpha\beta}\frac{dx_\alpha}{ds}dx_\beta.$$

The left-hand side is to be replaced by $\frac{d^2x_\mu}{ds^2}$,* so that we have

$$\frac{d^2x_\mu}{ds^2} + \Gamma^\mu_{\alpha\beta}\frac{dx_\alpha}{ds}\frac{dx_\beta}{ds} = 0. \tag{90}$$

We get the same line if we find the line which gives a stationary value to the integral

$$\int ds \text{ or } \int \sqrt{g_{\mu\nu}dx_\mu dx_\nu}$$

between two points (geodesic line).

* The direction vector at a neighbouring point of the curve results, by a parallel displacement along the line element ($dx_\beta$), from the direction vector of each point considered.

# THE GENERAL THEORY OF RELATIVITY (*Continued*)

WE are now in possession of the mathematical apparatus which is necessary to formulate the laws of the general theory of relativity. No attempt will be made in this presentation at systematic completeness, but single results and possibilities will be developed progressively from what is known and from the results obtained. Such a presentation is most suited to the present provisional state of our knowledge.

A material particle upon which no force acts moves, according to the principle of inertia, uniformly in a straight line. In the four-dimensional continuum of the special theory of relativity (with real time co-ordinate) this is a real straight line. The natural, that is, the simplest, generalization of the straight line which is meaningful in the system of concepts of the general (Riemannian) theory of invariants is that of the straightest, or geodesic, line. We shall accordingly have to assume, in the sense of the principle of equivalence, that the motion of a material particle, under the action only of inertia and gravitation, is described by the equation,

$$\frac{d^2x_\mu}{ds^2} + \Gamma^\mu_{\alpha\beta}\frac{dx_\alpha}{ds}\frac{dx_\beta}{ds} = 0. \qquad (90)$$

In fact, this equation reduces to that of a straight line if all the components, $\Gamma^\mu_{\alpha\nu}$, of the gravitational field vanish.

How are these equations connected with Newton's equations of motion? According to the special theory of relativity, the $g_{\mu\nu}$ as well as the $g^{\mu\nu}$, have the values, with respect to an inertial system (with real time co-ordinate and suitable choice of the sign of $ds^2$),

$$\left\{\begin{array}{cccc} -1 & 0 & 0 & 0 \\ 0 & -1 & 0 & 0 \\ 0 & 0 & -1 & 0 \\ 0 & 0 & 0 & 1 \end{array}\right. \tag{91}$$

The equations of motion then become

$$\frac{d^2x_\mu}{ds^2} = 0.$$

We shall call this the "first approximation" to the $g_{\mu\nu}$-field. In considering approximations it is often useful, as in the special theory of relativity, to use an imaginary $x_4$-co-ordinate, as then the $g_{\mu\nu}$, to the first approximation, assume the values

$$\left\{\begin{array}{cccc} -1 & 0 & 0 & 0 \\ 0 & -1 & 0 & 0 \\ 0 & 0 & -1 & 0 \\ 0 & 0 & 0 & -1 \end{array}\right. \tag{91a}$$

These values may be collected in the relation

$$g_{\mu\nu} = -\delta_{\nu\mu}.$$

To the second approximation we must then put

$$g_{\mu\nu} = -\delta_{\mu\nu} + \gamma_{\mu\nu} \tag{92}$$

where the $\gamma_{\mu\nu}$ are to be regarded as small of the first order.

Both terms of our equation of motion are then small of the first order. If we neglect terms which, relatively to these, are small of the first order, we have to put

$$ds^2 = -dx_\nu^2 = dl^2(1 - q^2) \tag{93}$$

$$\Gamma^\mu_{\alpha\beta} = -\delta_{\mu\sigma}\left[{\alpha\beta \atop \sigma}\right] = -\left[{\alpha\beta \atop \mu}\right] = \frac{1}{2}\left(\frac{\partial \gamma_{\alpha\beta}}{\partial x_\mu} - \frac{\partial \gamma_{\alpha\mu}}{\partial x_\beta} - \frac{\partial \gamma_{\beta\mu}}{\partial x_\alpha}\right). \tag{94}$$

We shall now introduce an approximation of a second kind. Let the velocity of the material particles be very small compared to that of light. Then $ds$ will be the same as the time differential, $dl$. Further, $\frac{dx_1}{ds}, \frac{dx_2}{ds}, \frac{dx_3}{ds}$ will vanish compared to $\frac{dx_4}{ds}$. We shall assume, in addition, that the gravitational field varies so little with the time that the derivatives of the $\gamma_{\mu\nu}$ by $x_4$ may be neglected. Then the equation of motion (for $\mu = 1, 2, 3$) reduces to

$$\frac{d^2 x_\mu}{dl^2} = \frac{\partial}{\partial x_\mu}\left(\frac{\gamma_{44}}{2}\right) \tag{90a}$$

This equation is identical with Newton's equation of motion for a material particle in a gravitational field, if we identify $\left(\frac{\gamma_{44}}{2}\right)$ with the potential of the gravitational field; whether or not this is allowable, naturally depends upon the field equations of gravitation, that is, it depends upon whether or not this quantity satisfies, to a first approximation, the same laws of the field as the gravitational potential in Newton's theory. A glance at (90) and (90a) shows that the $\Gamma^\mu_{\beta\alpha}$ actually do play the rôle of the intensity of the gravitational field. These quantities do not have a tensor character.

Equations (90) express the influence of inertia and gravitation upon the material particle. The unity of inertia and gravitation is formally expressed by the fact that the whole left-hand side of (90) has the character of a tensor (with respect to any transformation of co-ordinates), but the two terms taken separately do not have tensor character. In analogy with Newton's equations, the first term would be regarded as the expression for inertia, and the second as the expression for the gravitational force.

We must next attempt to find the laws of the gravitational field. For this purpose, Poisson's equation,

$$\Delta\phi = 4\pi K\rho$$

of the Newtonian theory must serve as a model. This equation has its foundation in the idea that the gravitational field arises from the density $\rho$ of ponderable matter. It must also be so in the general theory of relativity. But our investigations of the special theory of relativity have shown that in place of the scalar density of matter we have the tensor of energy per unit volume. In the latter is included not only the tensor of the energy of ponderable matter, but also that of the electromagnetic energy. We have seen, indeed, that in a more complete analysis the energy tensor can be regarded only as a provisional means of representing matter. In reality, matter consists of electrically charged particles, and is to be regarded itself as a part, in fact, the principal part, of the electromagnetic field. It is only the circumstance that we have no sufficient knowledge of the electromagnetic field of concentrated charges that compels us, provisionally, to leave undetermined in presenting the theory, the true form of this tensor. From this point of view it is at present appropriate to introduce a tensor $T_{\mu\nu}$ of the second rank of as yet unknown structure, which provisionally combines the energy density of the electromagnetic field and that of ponderable matter;

we shall denote this in the following as the "energy tensor of matter."

According to our previous results, the principles of momentum and energy are expressed by the statement that the divergence of this tensor vanishes (47c). In the general theory of relativity, we shall have to assume as valid the corresponding general co-variant equation. If $(T_{\mu\nu})$ denotes the co-variant energy tensor of matter, $\mathfrak{T}_\sigma^\nu$ the corresponding mixed tensor density, then, in accordance with (83), we must require that

$$0 = \frac{\partial \mathfrak{T}_\sigma^\alpha}{\partial x_\alpha} - \Gamma_{\sigma\beta}^\alpha \mathfrak{T}_\alpha^\beta \tag{95}$$

be satisfied. It must be remembered that besides the energy density of the matter there must also be given an energy density of the gravitational field, so that there can be no talk of principles of conservation of energy and momentum for matter alone. This is expressed mathematically by the presence of the second term in (95), which makes it impossible to conclude the existence of an integral equation of the form of (49). The gravitational field transfers energy and momentum to the "matter," in that it exerts forces upon it and gives it energy; this is expressed by the second term in (95).

If there is an analogue of Poisson's equation in the general theory of relativity, then this equation must be a tensor equation for the tensor $g_{\mu\nu}$ of the gravitational potential; the energy tensor of matter must appear on the right-hand side of this equation. On the left-hand side of the equation there must be a differential tensor in the $g_{\mu\nu}$. We have to find this differential tensor. It is completely determined by the following three conditions:

1. It may contain no differential coefficients of the $g_{\mu\nu}$ higher than the second.

2. It must be linear in these second differential coefficients.

3. Its divergence must vanish identically.

The first two of these conditions are naturally taken from Poisson's equation. Since it may be proved mathematically that all such differential tensors can be formed algebraically (i.e. without differentiation) from Riemann's tensor, our tensor must be of the form

$$R_{\mu\nu} + a g_{\mu\nu} R$$

in which $R_{\mu\nu}$ and $R$ are defined by (88) and (89) respectively. Further, it may be proved that the third condition requires $a$ to have the value $-\frac{1}{2}$. For the law of the gravitational field we therefore get the equation

$$R_{\mu\nu} - \tfrac{1}{2} g_{\mu\nu} R = -\kappa T_{\mu\nu}. \tag{96}$$

Equation (95) is a consequence of this equation. $\kappa$ denotes a constant, which is connected with the Newtonian gravitation constant.

In the following I shall indicate the features of the theory which are interesting from the point of view of physics, using as little as possible of the rather involved mathematical method. It must first be shown that the divergence of the left-hand side actually vanishes. The energy principle for matter may be expressed, by (83),

$$0 = \frac{\partial \mathfrak{T}_\sigma^\alpha}{\partial x_\alpha} - \Gamma_{\sigma\beta}^\alpha \mathfrak{T}_\alpha^\beta \tag{97}$$

in which

$$\mathfrak{T}_\sigma^\alpha = T_{\sigma\tau} g^{\tau\alpha} \sqrt{-g}.$$

The analogous operation, applied to the left-hand side of (96), will lead to an identity.

In the region surrounding each world-point there are systems of co-ordinates for which, choosing the $x_4$-coordinate imaginary, at the given point,

$$g_{\mu\nu} = g^{\mu\nu} = -\delta_{\mu\nu} \begin{cases} = -1 \text{ if } \mu = \nu \\ = 0 \text{ if } \mu \neq \nu, \end{cases}$$

and for which the first derivatives of the $g_{\mu\nu}$ and the $g^{\mu\nu}$ vanish. We shall verify the vanishing of the divergence of the left-hand side at this point. At this point the components $\Gamma^{\alpha}_{\sigma\beta}$ vanish, so that we have to prove the vanishing only of

$$\frac{\partial}{\partial x_\sigma}[\sqrt{-g}\, g^{\nu\sigma}(R_{\mu\nu} - \tfrac{1}{2}g_{\mu\nu}R)].$$

Introducing (88) and (70) into this expression, we see that the only terms that remain are those in which third derivatives of the $g_{\mu\nu}$ enter. Since the $g_{\mu\nu}$ are to be replaced by $-\delta_{\mu\nu}$, we obtain, finally, only a few terms which may easily be seen to cancel each other. Since the quantity that we have formed has a tensor character, its vanishing is proved for every other system of co-ordinates also, and naturally for every other four-dimensional point. The energy principle of matter (97) is thus a mathematical consequence of the field equations (96).

In order to learn whether the equations (96) are consistent with experience, we must, above all else, find out whether they lead to the Newtonian theory as a first approximation. For this purpose we must introduce various approximations into these equations. We already know that Euclidean geometry and the law of the constancy of the velocity of light are valid, to a certain approximation, in regions of a great extent, as in the planetary system. If, as in the special theory of relativity, we take the fourth co-ordinate imaginary, this means that we must put

$$g_{\mu\nu} = -\delta_{\mu\nu} + \gamma_{\mu\nu} \tag{98}$$

in which the $\gamma_{\mu\nu}$ are so small compared to 1 that we can neglect the higher powers of the $\gamma_{\mu\nu}$ and their derivatives. If we do this, we learn nothing about the structure of the gravitational field, or of metrical space of cosmical dimensions, but we do learn about the influence of neighbouring masses upon physical phenomena.

Before carrying through this approximation we shall transform (96). We multiply (96) by $g^{\mu\nu}$, summed over the $\mu$ and $\nu$; observing the relation which follows from the definition of the $g^{\mu\nu}$,

$$g_{\mu\nu}g^{\mu\nu} = 4$$

we obtain the equation

$$R = \kappa g^{\mu\nu}T_{\mu\nu} = \kappa T.$$

If we put this value of $R$ in (96) we obtain

$$R_{\mu\nu} = -\kappa(T_{\mu\nu} - \tfrac{1}{2}g_{\mu\nu}T) = -\kappa T^{*}_{\mu\nu}. \tag{96a}$$

When the approximation which has been mentioned is carried out, we obtain for the left-hand side,

$$-\frac{1}{2}\left(\frac{\partial^2\gamma_{\mu\nu}}{\partial x_\alpha^2} + \frac{\partial^2\gamma_{\alpha\alpha}}{\partial x_\mu \partial x_\nu} - \frac{\partial^2\gamma_{\mu\alpha}}{\partial x_\nu \partial x_\alpha} - \frac{\partial^2\gamma_{\nu\alpha}}{\partial x_\mu \partial x_\alpha}\right)$$

or

$$-\frac{1}{2}\frac{\partial^2\gamma_{\mu\nu}}{\partial x_\alpha^2} + \frac{1}{2}\frac{\partial}{\partial x_\nu}\left(\frac{\partial\gamma'_{\mu\alpha}}{\partial x_\alpha}\right) + \frac{1}{2}\frac{\partial}{\partial x_\mu}\left(\frac{\partial\gamma'_{\nu\alpha}}{\partial x_\alpha}\right)$$

in which has been put

$$\gamma'_{\mu\nu} = \gamma_{\mu\nu} - \tfrac{1}{2}\gamma_{\sigma\sigma}\delta_{\mu\nu}. \tag{99}$$

We must now note that equation (96) is valid for any system of co-ordinates. We have already specialized the system of co-ordinates in that we have chosen it so that within the region considered the $g_{\mu\nu}$ differ infinitely little from the constant values $-\delta_{\mu\nu}$. But this condition remains satisfied in any infinitesimal change of co-ordinates, so that there are still four conditions to which the $\gamma_{\mu\nu}$ may be subjected, provided these conditions do not conflict with the conditions for the order of magnitude of the $\gamma_{\mu\nu}$. We shall now assume that the system of co-ordinates is so chosen that the four relations—

$$0 = \frac{\partial \gamma'_{\mu\nu}}{\partial x_\nu} = \frac{\partial \gamma_{\mu\nu}}{\partial x_\nu} - \frac{1}{2}\frac{\partial \gamma_{\sigma\sigma}}{\partial x_\mu} \tag{100}$$

are satisfied. Then (96a) takes the form

$$\frac{\partial^2 \gamma_{\mu\nu}}{\partial {x_\alpha}^2} = 2\kappa T^*_{\mu\nu} \tag{96b}$$

These equations may be solved by the method, familiar in electrodynamics, of retarded potentials; we get, in an easily understood notation,

$$\gamma_{\mu\nu} = -\frac{\kappa}{2\pi}\int \frac{T^*_{\mu\nu}(x_0, y_0, z_0, t - r)}{r}\, dV_0. \tag{101}$$

In order to see in what sense this theory contains the Newtonian theory, we must consider in greater detail the energy tensor of matter. Considered phenomenologically, this energy tensor is composed of that of the electromagnetic field and of matter in the narrower sense. If we consider the different parts of this energy tensor with respect to their order of magnitude, it follows from the results of the special theory of relativity that the contribution of the electromagnetic field practically vanishes in comparison to that of ponderable matter. In our system of units, the energy of one gram of matter is equal to 1, compared to which the energy of the electric fields may be ignored, and also the energy of deformation of matter, and even the chemical energy. We get an approximation that is fully sufficient for our purpose if we put

$$(102) \qquad \begin{cases} T^{\mu\nu} = \sigma \dfrac{dx_\mu}{ds}\dfrac{dx_\nu}{ds} \\ ds^2 = g_{\mu\nu}dx_\mu dx_\nu \end{cases}$$

In this, $\sigma$ is the density at rest, that is, the density of the ponderable matter, in the ordinary sense, measured with the aid of a unit measuring rod, and referred to a Galilean system of co-ordinates moving with the matter.

We observe, further, that in the co-ordinates we have chosen, we shall make only a relatively small error if we replace the $g_{\mu\nu}$ by $-\delta_{\mu\nu}$, so that we put

$$(102a) \qquad ds^2 = -\sum dx_\mu{}^2.$$

The previous developments are valid however rapidly the masses which generate the field may move relatively to our chosen system of quasi-Galilean co-ordinates. But in astronomy we have to do with masses whose velocities, relatively to the co-ordinate system employed, are always small compared to the velocity of light, that is, small compared to 1, with our choice of the unit of time. We therefore get an approximation which is sufficient for nearly all practical purposes if in (101) we replace the retarded potential by the ordinary (non-retarded) potential, and if, for the masses which generate the field, we put

$$(103a) \qquad \frac{dx_1}{ds} = \frac{dx_2}{ds} = \frac{dx_3}{ds} = 0, \ \frac{dx_4}{ds} = \frac{\sqrt{-1}\,dl}{dl} = \sqrt{-1}.$$

Then we get for $T^{\mu\nu}$ and $T_{\mu\nu}$ the values

$$
(104) \qquad \left\{ \begin{array}{cccc} 0 & 0 & 0 & 0 \\ 0 & 0 & 0 & 0 \\ 0 & 0 & 0 & 0 \\ 0 & 0 & 0 & -\sigma \end{array} \right.
$$

For $T$ we get the value $\sigma$, and, finally, for $T^*_{\mu\nu}$ the values,

$$
(104a) \qquad \left\{ \begin{array}{cccc} \frac{\sigma}{2} & 0 & 0 & 0 \\ 0 & \frac{\sigma}{2} & 0 & 0 \\ 0 & 0 & \frac{\sigma}{2} & 0 \\ 0 & 0 & 0 & -\frac{\sigma}{2} \end{array} \right.
$$

We thus get, from (101),

$$
(101a) \qquad \left\{ \begin{aligned} \gamma_{11} = \gamma_{22} = \gamma_{33} &= -\frac{\kappa}{4\pi} \int \frac{\sigma dV_0}{r} \\ \gamma_{44} &= +\frac{\kappa}{4\pi} \int \frac{\sigma dV_0}{r} \end{aligned} \right.
$$

while at the other $\gamma_{\mu\nu}$ vanish. The last of these equations, in connexion with equation (90a), contains Newton's theory of gravitation. If we replace $l$ by $ct$ we get

$$
(90b) \qquad \left\{ \frac{d^2 x_\mu}{dt^2} = \frac{\kappa c^2}{8\pi} \frac{\partial}{\partial x_\mu} \int \frac{\sigma dV_0}{r}. \right.
$$

We see that the Newtonian gravitation constant $K$, is connected with the constant $\kappa$ that enters into our field equations by the relation

$$
(105) \qquad K = \frac{\kappa c^2}{8\pi}.
$$

From the known numerical value of $K$, it therefore follows that

$$(105a) \quad \kappa = \frac{8\pi K}{c^2} = \frac{8\pi \cdot 6{\cdot}67 \cdot 10^{-8}}{9 \cdot 10^{20}} = 1{\cdot}86 \cdot 10^{-27}.$$

From (101) we see that even in the first approximation the structure of the gravitational field differs fundamentally from that which is consistent with the Newtonian theory; this difference lies in the fact that the gravitational potential has the character of a tensor and not a scalar. This was not recognized in the past because only the component $g_{44}$, to a first approximation, enters the equations of motion of material particles.

In order now to be able to judge the behaviour of measuring rods and clocks from our results, we must observe the following. According to the principle of equivalence, the metrical relations of the Euclidean geometry are valid relatively to a Cartesian system of reference of infinitely small dimensions, and in a suitable state of motion (freely falling, and without rotation). We can make the same statement for local systems of co-ordinates which, relatively to these, have small accelerations, and therefore for such systems of co-ordinates as are at rest relatively to the one we have selected. For such a local system, we have, for two neighbouring point events,

$$ds^2 = -dX_1^2 - dX_2^2 - dX_3^2 + dT^2 = -dS^2 + dT^2$$

where $dS$ is measured directly by a measuring rod and $dT$ by a clock at rest relatively to the system: these are the naturally measured lengths and times. Since $ds^2$, on the other hand, is known in terms of the co-ordinates $x_\nu$ employed in finite regions, in the form

$$ds^2 = g_{\mu\nu} dx_\mu dx_\nu$$

we have the possibility of getting the relation between naturally measured lengths and times, on the one hand, and the corresponding differences of co-ordinates, on the other hand. As the division into space and time is in agreement with respect to the two systems of co-ordinates, so when we equate the two expressions for $ds^2$ we get two relations. If, by (101a), we put

$$ds^2 = -\left(1 + \frac{\kappa}{4\pi}\int\frac{\sigma dV_0}{r}\right)(dx_1^2 + dx_2^2 + dx_3^2) + \left(1 - \frac{\kappa}{4\pi}\int\frac{\sigma dV_0}{r}\right)dl^2$$

we obtain, to a sufficiently close approximation,

$$\left\{\begin{aligned}&\sqrt{dX_1^2 + dX_2^2 + dX_3^2}\\ &\qquad = \left(1 + \frac{\kappa}{8\pi}\int\frac{\sigma dV_0}{r}\right)\sqrt{dx_1^2 + dx_2^2 + dx_3^2}\\ &dT = \left(1 - \frac{\kappa}{8\pi}\int\frac{\sigma dV_0}{r}\right)dl.\end{aligned}\right. \tag{106}$$

The unit measuring rod has therefore the coordinate length,

$$1 - \frac{\kappa}{8\pi}\int\frac{\sigma dV_0}{r}$$

in respect to the system of co-ordinates we have selected. The particular system of co-ordinates we have selected insures that this length shall depend only upon the place, and not upon the direction. If we had chosen a different system of co-ordinates this would not be so. But however we may choose a system of co-ordinates, the laws of configuration of rigid rods do not agree with those of Euclidean geometry; in other words, we cannot choose any system of co-ordinates so that the co-ordinate differences, $\Delta x_1$,

$\Delta x_2$, $\Delta x_3$, corresponding to the ends of a unit measuring rod, oriented in any way, shall always satisfy the relation $\Delta x_1^2 + \Delta x_2^2 + \Delta x_3^2 = 1$. In this sense space is not Euclidean, but "curved." It follows from the second of the relations above that the interval between two beats of the unit clock ($dT = 1$) corresponds to the "time"

$$1 + \frac{\kappa}{8\pi} \int \frac{\sigma dV_0}{r}$$

in the unit used in our system of co-ordinates. The rate of a clock is accordingly slower the greater is the mass of the ponderable matter in its neighbourhood. We therefore conclude that spectral lines which are produced on the sun's surface will be displaced towards the red, compared to the corresponding lines produced on the earth, by about $2 \,.\, 10^{-6}$ of their wave-lengths. At first, this important consequence of the theory appeared to conflict with experiment; but results obtained during the past years seem to make the existence of this effect more and more probable, and it can hardly be doubted that this consequence of the theory will be confirmed within the next years.

Another important consequence of the theory, which can be tested experimentally, has to do with the path of rays of light. In the general theory of relativity also the velocity of light is everywhere the same, relatively to a local inertial system. This velocity is unity in our natural measure of time. The law of the propagation of light in general co-ordinates is therefore, according to the general theory of relativity, characterized, by the equation

$$ds^2 = 0.$$

To within the approximation which we are using, and in the system of co-ordinates which we have selected, the velocity of light is characterized, according to (106), by the equation

$$\left(1+\frac{\kappa}{4\pi}\int\frac{\sigma dV_0}{r}\right)(dx_1^2+dx_2^2+dx_3^2)$$
$$=\left(1-\frac{\kappa}{4\pi}\int\frac{\sigma dV_0}{r}\right)dl^2.$$

The velocity of light $L$, is therefore expressed in our co-ordinates by

$$\frac{\sqrt{dx_1^2+dx_2^2+dx_3^2}}{dl}=1-\frac{\kappa}{4\pi}\int\frac{\sigma dV_0}{r} \tag{107}$$

We can therefore draw the conclusion from this, that a ray of light passing near a large mass is deflected. If we imagine the sun, of mass $M$, concentrated at the origin of our system of co-ordinates, then a ray of light, travelling parallel to the $x_3$-axis, in the $x_1-x_2$ plane, at a distance $\Delta$ from the origin, will be deflected, in all, by an amount

$$\alpha=\int_{-\infty}^{+\infty}\frac{1}{L}\frac{\partial L}{\partial x_1}dx_3$$

towards the sun. On performing the integration we get

$$\alpha=\frac{\kappa M}{2\pi\Delta}. \tag{108}$$

The existence of this deflection, which amounts to 1.7″ for $\Delta$ equal to the radius of the sun, was confirmed, with remarkable accuracy, by the English Solar Eclipse Expedition in 1919, and most careful preparations have been made to get more exact observational data at the solar eclipse in 1922. It should be noted that this result, also, of the theory is not influenced by our arbitrary choice of a system of co-ordinates.

This is the place to speak of the third consequence of the theory which can be tested by observation, namely, that which concerns the motion of the perihelion of the planet Mercury. The secular changes in the planetary orbits are known with such accuracy that the approximation we have been using is no longer sufficient for a comparison of theory and observation. It is necessary to go back to the general field equations (96). To solve this problem I made use of the method of successive approximations. Since then, however, the problem of the central symmetrical statical gravitational field has been completely solved by Schwarzschild and others; the derivation given by H. Weyl in his book, "Raum-Zeit-Materie," is particularly elegant. The calculation can be simplified somewhat if we do not go back directly to the equation (96), but base it upon a principle of variation that is equivalent to this equation. I shall indicate the procedure only in so far as is necessary for understanding the method.

In the case of a statical field, $ds^2$ must have the form

$$\left\{\begin{aligned} ds^2 &= -d\sigma^2 + f^2 dx_4{}^2 \\ d\sigma^2 &= \sum_{1-3} \gamma_{\alpha\beta} dx_\alpha dx_\beta \end{aligned}\right. \tag{109}$$

where the summation on the right-hand side of the last equation is to be extended over the space variables only, The central symmetry of the field requires the $\gamma_{\mu\nu}$ to be of the form,

$$\gamma_{\alpha\beta} = \mu\delta_{\alpha\beta} + \lambda x_\alpha x_\beta \tag{110}$$

$f^2$, $\mu$ and $\lambda$ are functions of $r = \sqrt{x_1{}^2 + x_2{}^2 + x_3{}^2}$ only. One of these three functions can be chosen arbitrarily, because our system of co-ordinates is, *a priori*, completely arbitrary; for by a substitution

$$x'_4 = x_4$$
$$x'_\alpha = F(r)x_\alpha$$

we can always insure that one of these three functions shall be an assigned function of $r'$. In place of (110) we can therefore put, without limiting the generality,

$$\gamma_{\alpha\beta} = \delta_{\alpha\beta} + \lambda x_\alpha x_\beta. \tag{110a}$$

In this way the $g_{\mu\nu}$ are expressed in terms of the two quantities $\lambda$ and $f$. These are to be determined as functions of $r$, by introducing them into equation (96), after first calculating the $\Gamma^\sigma_{\mu\nu}$ from (109) and (110a). We have

$$\left\{\begin{aligned} \Gamma^\sigma_{\alpha\beta} &= \tfrac{1}{2}\frac{x_\sigma}{r}\cdot\frac{\lambda' x_\alpha x_\beta + 2\lambda r\delta_{\alpha\beta}}{1+\lambda r^2} \quad (\text{for } \alpha, \beta, \sigma = 1, 2, 3) \\ \Gamma^4_{44} &= \Gamma^\alpha_{4\beta} = \Gamma^4_{\alpha\beta} = 0 \quad (\text{for } \alpha, \beta = 1, 2, 3) \\ \Gamma^4_{4\alpha} &= \tfrac{1}{2}f^{-2}\frac{\partial f^2}{\partial x_\alpha}, \qquad \Gamma^\alpha_{44} = -\tfrac{1}{2}g^{\alpha\beta}\frac{\partial f^2}{\partial x_\beta} \end{aligned}\right. \tag{110b}$$

With the help of these results, the field equations furnish Schwarzschild's solution:

$$ds^2 = \left(1 - \frac{A}{r}\right)dl^2 - \left[\frac{dr^2}{1 - \frac{A}{r}} + r^2(\sin^2\theta d\phi^2 + d\theta^2)\right] \tag{109a}$$

in which we have put

$$\left\{\begin{aligned} x_4 &= l \\ x_1 &= r\sin\theta\sin\phi \\ x_2 &= r\sin\theta\cos\phi \\ x_3 &= r\cos\theta \\ A &= \frac{\kappa M}{4\pi} \end{aligned}\right. \tag{109b}$$

$M$ denotes the sun's mass, centrally symmetrically placed about the origin of co-ordinates; the solution (109a) is valid only outside of this mass, where all the $T_{\mu\nu}$ vanish. If the motion of the planet takes place in the $x_1 - x_2$ plane then we must replace (109a) by

$$ds^2 = \left(1 - \frac{A}{r}\right) dl^2 - \frac{dr^2}{1 - \frac{A}{r}} - r^2 d\phi^2. \tag{109c}$$

The calculation of the planetary motion depends upon equation (90). From the first of equations (110b) and (90) we get, for the indices 1, 2, 3,

$$\frac{d}{ds}\left(x_\alpha \frac{dx_\beta}{ds} - x_\beta \frac{dx_\alpha}{ds}\right) = 0$$

or, if we integrate, and express the result in polar coordinates,

$$r^2 \frac{d\phi}{ds} = \text{constant}. \tag{111}$$

From (90), for $\mu = 4$, we get

$$0 = \frac{d^2 l}{ds^2} + \frac{1}{f^2} \frac{\partial f^2}{\partial x_\alpha} \frac{dx_\alpha}{ds} \frac{dl}{ds} = \frac{d^2 l}{ds^2} + \frac{1}{f^2} \frac{df^2}{ds} \frac{dl}{ds}.$$

From this, after multiplication by $f^2$ and integration, we have

$$f^2 \frac{dl}{ds} = \text{constant}. \tag{112}$$

In (109c), (111) and (112) we have three equations between the four variables $s$, $r$, $l$ and $\phi$, from which the motion of the planet may be calculated in the same way as in classical mechanics. The most important result we get from this is a secular rotation of the elliptic orbit of the planet in the same sense as the revolution of the planet, amounting in radians per revolution to

$$\frac{24\pi^3 a^2}{(1 - e^2)\, c^2 T^2} \tag{113}$$

where

$a$ = the semi-major axis of the planetary orbit in centimetres.
$e$ = the numerical eccentricity.
$c = 3 \,.\, 10^{+10}$, the velocity of the light *in vacuo*.
$T$ = the period of revolution in seconds.

This expression furnishes the explanation of the motion of the perihelion of the planet Mercury, which has been known for a hundred years (since Leverrier), and for which theoretical astronomy has hitherto been unable satisfactorily to account.

There is no difficulty in expressing Maxwell's theory of the electromagnetic field in terms of the general theory of relativity; this is done by application of the tensor formation (81), (82) and (77). Let $\phi_\mu$ be a tensor of the first rank, to be interpreted as an electromagnetic 4-potential; then as electromagnetic field tensor may be defined by the relations,

$$\phi_{\mu\nu} = \frac{\partial \phi_\mu}{\partial x_\nu} - \frac{\partial \phi_\nu}{\partial x_\mu}. \tag{114}$$

The second of Maxwell's systems of equations is then defined by the tensor equation, resulting from this,

$$\frac{\partial \phi_{\mu\nu}}{\partial x_\rho} + \frac{\partial \phi_{\nu\rho}}{\partial x_\mu} + \frac{\partial \phi_{\rho\mu}}{\partial x_\nu} = 0 \tag{114a}$$

and the first of Maxwell's systems of equations is defined by the tensor-density relation

$$\frac{\partial \mathfrak{F}^{\mu\nu}}{\partial x_\nu} = \mathfrak{J}^\mu \tag{115}$$

in which

$$\mathfrak{F}^{\mu\nu} = \sqrt{-g}\, g^{\mu\sigma} g^{\nu\tau} \phi_{\sigma\tau}$$

$$\mathfrak{J}^\mu = \sqrt{-g}\, \rho \frac{dx_\mu}{ds}.$$

If we introduce the energy tensor of the electromagnetic field into the right-hand side of (96), we obtain (115), for the special case $\mathfrak{J}^\mu = 0$, as a consequence of (96) by taking the divergence. This inclusion of the theory of electricity in the scheme of the general theory of relativity has been considered arbitrary and unsatisfactory by many theoreticians. Nor can we in this way understand the equilibrium of the electricity which constitutes the elementary electrically charged particles. A theory in which the gravitational field and the electromagnetic field do not enter as logically distinct structures would be much preferable. H. Weyl, and recently Th. Kaluza, have put forward ingenious ideas along this direction; but concerning them, I am convinced that they do not bring us nearer to the true solution of the fundamental problem. I shall not go into this further, but shall give a brief discussion of the so-called cosmological problem, for without this, the considerations regarding the general theory of relativity would, in a certain sense, remain unsatisfactory.

Our previous considerations, based upon the field equations (96), had for a foundation the conception that space on the whole is Galilean-Euclidean, and that this character is disturbed only by masses embedded in it. This conception was certainly justified as long as we were dealing with spaces of the order of magnitude of those that astronomy has mostly to do with. But whether portions of the universe, however large they may be, are quasi-Euclidean, is a wholly different question. We can make this clear by using an example from the theory of surfaces which we have employed many times. If a certain portion of a surface is practically plane, it does not at all follow that the whole surface has the form of a plane; the surface might just as well be a sphere of sufficiently large radius. The question as to whether the universe as a whole is non-Euclidean was much discussed from the geometrical point of view before the development of the theory of relativity. But with the theory of relativity, this problem has entered upon a new stage, for according to this theory the geometrical properties of bodies are not independent, but depend upon the distribution of masses.

If the universe were quasi-Euclidean, then Mach was wholly wrong in his thought that inertia, as well as gravitation, depends upon a kind of mutual action between bodies. For in this case, for a suitably selected system of co-ordinates, the $g_{\mu\nu}$ would be constant at infinity, as they are in the special theory of relativity, while within finite regions the $g_{\mu\nu}$ would differ from these constant values by small amounts only, for a suitable choice co-ordinates, as a result of the influence of the masses in finite regions. The physical properties of space would not then be wholly independent, that is, uninfluenced by matter, but in the main they would be, and only in small measure conditioned by matter. Such a dualistic conception is even in itself not satisfactory; there are, however, some important physical arguments against it, which we shall consider.

The hypothesis that the universe is infinite and Euclidean at infinity, is, from the relativistic point of view, a complicated hypothesis. In the language of the general theory of relativity it demands that the Riemann tensor of the fourth rank, $R_{iklm}$, shall vanish at infinity, which furnishes twenty independent conditions, while only ten curvature components, $R_{\mu\nu}$, enter into the laws of the gravitational field. It is certainly unsatisfactory to postulate such a far-reaching limitation without any physical basis for it.

But in the second place, the theory of relativity makes it appear probable that Mach was on the right road in his thought that inertia depends upon a mutual action of matter. For we shall show in the following that, according to our equations, inert masses do act upon each other in the sense of the relativity of inertia, even if only very feebly. What is to be expected along the line of Mach's thought?

1. The inertia of a body must increase when ponderable masses are piled up in its neighbourhood.
2. A body must experience an accelerating force when neighbouring masses are accelerated, and, in fact, the force must be in the same direction as that acceleration.
3. A rotating hollow body must generate inside of itself a "Coriolis field," which deflects moving bodies in the sense of the rotation, and a radial centrifugal field as well.

We shall now show that these three effects, which are to be expected in accordance with Mach's ideas, are actually present according to our theory, although their magnitude is so small that confirmation of them by laboratory experiments is not to be thought of. For this purpose we shall go back to the equations of motion of a material particle (90), and carry the approximations somewhat further than was done in equation (90a).

First, we consider $\gamma_{44}$ as small of the first order. The square of the velocity of masses moving under the influence of the gravitational force is of the same order, according to the energy equation. It is therefore logical to regard the velocities of the material particles we are considering, as well as the velocities of the masses which generate the field, as small, of the order $\frac{1}{2}$. We shall now carry out the approximation in the equations that arise from the field equations (101) and the equations of motion (90) so far as to consider terms, in the second member of (90), that are linear in those velocities. Further, we shall not put $ds$ and $dl$ equal to each other, but, corresponding to the higher approximation, we shall put

$$ds = \sqrt{g_{44}}\, dl = \left(1 - \frac{\gamma_{44}}{2}\right) dl.$$

From (90) we obtain, at first,

$$\frac{d}{dl}\left[\left(1 + \frac{\gamma_{44}}{2}\right)\frac{dx_\mu}{dl}\right] = -\Gamma^\mu_{\alpha\beta}\frac{dx_\alpha}{dl}\frac{dx_\beta}{dl}\left(1 + \frac{\gamma_{44}}{2}\right). \tag{116}$$

From (101) we get, to the approximation sought for,

$$\left.\begin{aligned} -\gamma_{11} = -\gamma_{22} = -\gamma_{33} = \gamma_{44} &= \frac{\kappa}{4\pi}\int\frac{\sigma dV_0}{r} \\ \gamma_{4\alpha} &= -\frac{i\kappa}{2\pi}\int\frac{\sigma\frac{dx_\alpha}{ds}dV_0}{r} \\ \gamma_{\alpha\beta} &= 0 \end{aligned}\right\} \tag{117}$$

in which, in (117), $\alpha$ and $\beta$ denote the space indices only.

On the right-hand side of (116) we can replace $1 + \frac{\gamma_{44}}{2}$ by 1 and $-\Gamma^{\alpha\beta}_\mu$ by $\left[{\alpha\beta \atop \mu}\right]$. It is easy to see, in addition, that to this degree of approximation we must put

$$\begin{bmatrix} 44 \\ \mu \end{bmatrix} = -\tfrac{1}{2}\frac{\partial \gamma_{44}}{\partial x_\mu} + \frac{\partial \gamma_{4\mu}}{\partial x_4}$$

$$\begin{bmatrix} \alpha 4 \\ \mu \end{bmatrix} = \tfrac{1}{2}\left(\frac{\partial \gamma_{4\mu}}{\partial x_\alpha} - \frac{\partial \gamma_{4\alpha}}{\partial x_\mu}\right)$$

$$\begin{bmatrix} \alpha\beta \\ \mu \end{bmatrix} = 0$$

in which $\alpha$, $\beta$ and $\mu$ denote space indices. We therefore obtain from (116), in the usual vector notation,

$$\left.\begin{aligned} \frac{d}{dl}[(1+\bar{\sigma})\mathbf{v}] &= \operatorname{grad}\bar{\sigma} + \frac{\partial \mathfrak{A}}{\partial l} + [\operatorname{curl}\mathfrak{A}, \mathbf{v}] \\ \bar{\sigma} &= \frac{\kappa}{8\pi}\int \frac{\sigma dV_0}{r} \\ \mathfrak{A} &= \frac{\kappa}{2\pi}\int \frac{\sigma \frac{dx_\alpha}{dl} dV_0}{r} \end{aligned}\right\} \tag{118}$$

The equations of motion, (118), show now, in fact, that

1. The inert mass is proportional to $1 + \bar{\sigma}$, and therefore increases when ponderable masses approach the test body.
2. There is an inductive action of accelerated masses, of the same sign, upon the test body. This is the term $\frac{\partial \mathfrak{A}}{\partial l}$.
3. A material particle, moving perpendicularly to the axis of rotation inside a rotating hollow body, is deflected in the sense of the rotation (Coriolis field). The centrifugal action, mentioned above, inside a rotating hollow body, also follows from the theory, as has been shown by Thirring.*

* That the centrifugal action must be inseparably connected with the existence of the Coriolis field may be recognized, even without calculation, in the special case of a co-ordinate system rotating uniformly relatively to an inertial system; our general co-variant equations naturally must apply to such a case.

Although all of these effects are inaccessible to experiment, because $\kappa$ is so small, nevertheless they certainly exist according to the general theory of relativity. We must see in them a strong support for Mach's ideas as to the relativity of all inertial actions. If we think these ideas consistently through to the end we must expect the *whole* inertia, that is, the *whole* $g_{\mu\nu}$-field, to be determined by the matter of the universe, and not mainly by the boundary conditions at infinity.

For a satisfactory conception of the $g_{\mu\nu}$-field of cosmical dimensions, the fact seems to be of significance that the relative velocity of the stars is small compared to the velocity of light. It follows from this that, with a suitable choice of co-ordinates, $g_{44}$ is nearly constant in the universe, at least, in that part of the universe in which there is matter. The assumption appears natural, moreover, that there are stars in all parts of the universe, so that we may well assume that the inconstancy of $g_{44}$ depends only upon the circumstance that matter is not distributed continuously, but is concentrated in single celestial bodies and systems of bodies. If we are willing to ignore these more local non-uniformities of the density of matter and of the $g_{\mu\nu}$-field, in order to learn something of the geometrical properties of the universe as a whole, it appears natural to substitute for the actual distribution of masses a continuous distribution, and furthermore to assign to this distribution a uniform density $\sigma$. In this imagined universe all points with space directions will be geometrically equivalent; with respect to its space extension it will have a constant curvature, and will be cylindrical with respect to its $x_4$-co-ordinate. The possibility seems to be particularly satisfying that the universe is spatially bounded and thus, in accordance with our assumption of the constancy of $\sigma$, is of constant curvature, being either spherical or elliptical; for then the boundary conditions at infinity which are so inconvenient from the standpoint of the general theory of relativity, may be replaced by the much more natural conditions for a closed space.

According to what has been said, we are to put

$$ds^2 = dx_4^2 - \gamma_{\mu\nu} dx_\mu dx_\nu \tag{119}$$

in which the indices $\mu$ and $\nu$ run from 1 to 3 only. The $\gamma_{\mu\nu}$ will be such functions of $x_1$, $x_2$, $x_3$ as correspond to a three-dimensional continuum of constant positive curvature. We must now investigate whether such an assumption can satisfy the field equations of gravitation.

In order to be able to investigate this, we must first find what differential conditions the three-dimensional manifold of constant curvature satisfies. A spherical manifold of three dimensions, embedded in a Euclidean continuum of four dimensions,* is given by the equations

$$x_1^2 + x_2^2 + x_3^2 + x_4^2 = a^2$$
$$dx_1^2 + dx_2^2 + dx_3^2 + dx_4^2 = ds^2.$$

By eliminating $x_4$, we get

$$ds^2 = dx_1^2 + dx_2^2 + dx_3^2 + \frac{(x_1 dx_1 + x_2 dx_2 + x_3 dx_3)^2}{a^2 - x_1^2 - x_2^2 - x_3^2}.$$

Neglecting terms of the third and higher degrees in the $x_\nu$, we can put, in the neighbourhood of the origin of co-ordinates,

$$ds^2 = \left(\delta_{\mu\nu} + \frac{x_\mu x_\nu}{a^2}\right) dx_\mu dx_\nu.$$

* The aid of a fourth space dimension has naturally no significance except that of a mathematical artifice.

Inside the brackets are the $g_{\mu\nu}$ of the manifold in the neighbourhood of the origin. Since the first derivatives of the $g_{\mu\nu}$, and therefore also the $\Gamma^{\sigma}_{\mu\nu}$, vanish at the origin, the calculation of the $R_{\mu\nu}$ for this manifold, by (88), is very simple at the origin. We have

$$R_{\mu\nu} = -\frac{2}{a^2}\delta_{\mu\nu} = -\frac{2}{a^2}g_{\mu\nu}.$$

Since the relation $R_{\mu\nu} = -\frac{2}{a^2}g_{\mu\nu}$ is generally co-variant, and since all points of the manifold are geometrically equivalent, this relation holds for every system of co-ordinates, and everywhere in the manifold. In order to avoid confusion with the four-dimensional continuum, we shall, in the following, designate quantities that refer to the three-dimensional continuum by Greek letters, and put

$$\text{(120)} \qquad \mathrm{P}_{\mu\nu} = -\frac{2}{a^2}\gamma_{\mu\nu}.$$

We now proceed to apply the field equations (96) to our special case. From (119) we get for the four-dimensional manifold,

$$\text{(121)} \qquad \begin{cases} R_{\mu\nu} = \mathrm{P}_{\mu\nu} \text{ for the indices 1 to 3} \\ R_{14} = R_{24} = R_{34} = R_{44} = 0 \end{cases}$$

For the right-hand side of (96) we have to consider the energy tensor for matter distributed like a cloud of dust. According to what has gone before we must therefore put

$$T^{\mu\nu} = \sigma\frac{dx_\mu}{ds}\frac{dx_\nu}{ds}$$

specialized for the case of rest. But in addition, we shall add a pressure term that may be physically established as follows. Matter consists of electrically charged particles. On the basis of Maxwell's theory these cannot be conceived of as electromagnetic fields free from singularities. In order to be consistent with the facts, it is necessary to introduce energy terms, not contained in Maxwell's theory, so that the single electric particles may hold together in spite of the mutual repulsions between their elements, charged with electricity of one sign. For the sake of consistency with this fact, Poincaré has assumed a pressure to exist inside these particles which balances the electrostatic repulsion. It cannot, however, be asserted that this pressure vanishes outside the particles. We shall be consistent with this circumstance if, in our phenomenological presentation, we add a pressure term. This must not, however, be confused with a hydrodynamical pressure, as it serves only for the energetic presentation of the dynamical relations inside matter. Accordingly we put

$$T_{\mu\nu} = g_{\mu\alpha} g_{\nu\beta} \sigma \frac{dx_\alpha}{ds} \frac{dx_\beta}{ds} - g_{\mu\nu} p. \tag{122}$$

In our special case we have, therefore, to put

$$\begin{aligned} T_{\mu\nu} &= \gamma_{\mu\nu} p \text{ (for } \mu \text{ and } \nu \text{ from 1 to 3)} \\ T_{44} &= \sigma - p \\ T &= -\gamma^{\mu\nu}\gamma_{\mu\nu} p + \sigma - p = \sigma - 4p. \end{aligned}$$

Observing that the field equation (96) may be written in the form

$$R_{\mu\nu} = -\kappa(T_{\mu\nu} - \tfrac{1}{2} g_{\mu\nu} T)$$

we get from (96) the equations,

$$\begin{aligned} +\frac{2}{a^2}\gamma_{\mu\nu} &= \kappa\left(\frac{\sigma}{2} - p\right)\gamma_{\mu\nu} \\ 0 &= -\kappa\left(\frac{\sigma}{2} + p\right). \end{aligned}$$

From this follows

$$(123) \qquad \begin{cases} p = -\dfrac{\sigma}{2} \\ a = \sqrt{\dfrac{2}{\kappa\sigma}} \end{cases}$$

If the universe is quasi-Euclidean, and its radius of curvature therefore infinite, then $\sigma$ would vanish. But it is improbable that the mean density of matter in the universe is actually zero; this is our third argument against the assumption that the universe is quasi-Euclidean. Nor does it seem possible that our hypothetical pressure can vanish; the physical nature of this pressure can be appreciated only after we have a better theoretical knowledge of the electromagnetic field. According to the second of equations (123) the radius, $a$, of the universe is determined in terms of the total mass, $M$, of matter, by the equation

$$(124) \qquad a = \frac{M\kappa}{4\pi^2}.$$

The complete dependence of the geometrical upon the physical properties becomes clearly apparent by means of this equation.

Thus we may present the following arguments against the conception of a space-infinite, and for the conception of a space-bounded, or closed, universe:—

1. From the standpoint of the theory of relativity, to postulate a closed universe is very much simpler than to postulate the corresponding boundary condition at infinity of the quasi-Euclidean structure of the universe.

2. The idea that Mach expressed, that inertia depends upon the mutual action of bodies, is contained, to a first approximation, in the equations of the theory of relativity;

it follows from these equations that inertia depends, at least in part, upon mutual actions between masses. Thereby Mach's idea gains in probability, as it is an unsatisfactory assumption to make that inertia depends in part upon mutual actions, and in part upon an independent property of space. But this idea of Mach's corresponds only to a finite universe, bounded in space, and not to a quasi-Euclidean, infinite universe. From the standpoint of epistemology it is more satisfying to have the mechanical properties of space completely determined by matter, and this is the case only in a closed universe.

3. An infinite universe is possible only if the mean density of matter in the universe vanishes. Although such an assumption is logically possible, it is less probable than the assumption that there is a finite mean density of matter in the universe.

# APPENDIX
# FOR THE SECOND EDITION

## ON THE "COSMOLOGIC PROBLEM"

SINCE the first edition of this little book some advances have been made in the theory of relativity. Some of these we shall mention here only briefly:

The first step forward is the conclusive demonstration of the existence of the red shift of the spectral lines by the (negative) gravitational potential of the place of origin (see p. 92). This demonstration was made possible by the discovery of so-called "dwarf stars" whose average density exceeds that of water by a factor of the order $10^4$. For such a star (e.g. the faint companion of Sirius), whose mass and radius can be determined,* this red shift was expected, by the theory, to be about 20 times as large as for the sun, and indeed it was demonstrated to be within the expected range.

A second step forward, which will be mentioned briefly, concerns the law of motion of a gravitating body. In the initial formulation of the theory the law of motion for a gravitating particle was introduced as an independent fundamental assumption in addition to the field law of gravitation—see Eq. 90 which asserts that a gravitating particle moves in a geodesic line. This constitutes a

* The mass is derived from the reaction on Sirius by spectroscopic means, using the Newtonian laws; the radius is derived from the total lightness and from the intensity of radiation per unit area, which may be derived from the temperature of its radiation.

hypothetic translation of Galileo's law of inertia to the case of the existence of "genuine" gravitational fields. It has been shown that this law of motion—generalized to the case of arbitrarily large gravitating masses—can be derived from the field-equations of empty space alone. According to this derivation the law of motion is implied by the condition that the field be singular nowhere outside its generating mass points.

A third step forward, concerning the so-called "cosmologic problem," will be considered here in detail, in part because of its basic importance, partly also because the discussion of these questions is by no means concluded. I feel urged toward a more exact discussion also by the fact that I cannot escape the impression that in the present treatment of this problem the most important basic points of view are not sufficiently stressed.

The problem can be formulated roughly thus: On account of our observations on fixed stars we are sufficiently convinced that the system of fixed stars does not in the main resemble an island which floats in infinite empty space, and that there does not exist anything like a center of gravity of the total amount of existing matter. Rather, we feel urged toward the conviction that there exists an average density of matter in space which differs from zero.

Hence the question arises: Can this hypothesis, which is suggested by experience, be reconciled with the general theory of relativity?

First we have to formulate the problem more precisely. Let us consider a finite part of the universe which is large enough so that the average density of matter contained in it is an approximately continuous function of $(x_1, x_2, x_3, x_4)$. Such a subspace can be considered approximately as an inertial system (Minkowski space) to which we relate the motion of the stars. One can arrange it so that the mean velocity of matter relative to this system shall vanish in all directions. There remain the (almost random) motions of the individual stars, similar to the motions of the molecules of a gas. It is essential that the velocities of the stars are known by experience to be very small as compared to the velocity of light. It is therefore feasible for the moment to neglect this relative motion completely, and to consider the stars replaced by material dust without (random) motion of the particles against each other.

The above conditions are by no means sufficient to make the problem a definite one. The simplest and most radical specialization would be the condition: The (naturally measured) density, $\rho$ of matter is the same everywhere in (four-dimensional) space, the metric is, for a suitable choice of coordinates, independent of $x_4$ and homogeneous and isotropic with respect to $x_1$, $x_2$, $x_3$.

It is this case which I at first considered the most natural idealized description of physical space in the large; it is treated on pages 103–108 of this book. The objection to this solution is that one has to introduce a negative pressure, for which there exists no physical justification. In order to make that solution possible I originally introduced a new member into the equation instead of the above mentioned pressure, which is permissible from the point of view of relativity. The equations of gravitation thus enlarged were:

$$(1) \qquad (R_{ik} - \tfrac{1}{2}g_{ik}R) + \Lambda g_{ik} + \kappa T_{ik} = 0$$

where $\Lambda$ is a universal constant ("cosmologic constant"). The introduction of this second member constitutes a complication of the theory, which seriously reduces its logical simplicity. Its introduction can only be justified by the difficulty produced by the almost unavoidable introduction of a finite average density of matter. We may remark, by the way, that in Newton's theory there exists the same difficulty.

The mathematician Friedman found a way out of this dilemma.* His result then found a surprising confirmation by Hubble's discovery of the expansion of the stellar system (a red shift of the spectral lines which increases uniformly with distance). The following is essentially nothing but an exposition of Friedman's idea:

### FOUR-DIMENSIONAL SPACE WHICH IS ISOTROPIC WITH RESPECT TO THREE DIMENSIONS

We observe that the systems of stars, as seen by us, are spaced with approximately the same density in all directions. Thereby we are moved to the assumption that the *spatial* isotropy of the system would hold for all observers, for every place and every time of an observer who is at rest as compared with surrounding matter. On the other hand we no longer make the assumption that the average density of matter, for an observer who is at rest relative to neighboring matter, is constant with respect to time. With this we drop the assumption that the expression of the metric field is independent of time.

We now have to find a mathematical form for the condition that the universe, *spatially speaking*, is isotropic everywhere. Through every point $P$ of (four-dimensional) space there is the path of a particle (which in the following will be called "geodesic" for short). Let $P$ and $Q$ be two

* He showed that it is possible, according to the field equations, to have a finite density in the whole (three-dimensional) space, without enlarging these field equations *ad hoc*. Zeitschr. f. Phys. 10 (1922).

infinitesimally near points of such a geodesic. We shall then have to demand that the expression of the field shall be invariant relative to any rotation of the coordinate system keeping $P$ and $Q$ fixed. This will be valid for any element of any geodesic.*

The condition of the above invariance implies that the entire geodesic lies on the axis of rotation and that its points remain invariant under rotation of the coordinate system. This means that the solution shall be invariant with respect to all rotations of the coordinate system around the triple infinity of geodesics.

For the sake of brevity I will not go into the deductive derivation of the solution of this problem. It seems intuitively evident, however, for the three-dimensional space that a metric which is invariant under rotations around a double infinity of lines will be essentially of the type of central symmetry (by suitable choice of coordinates), where the axes of rotations are the radial straight lines, which by reasons of symmetry are geodesics. The surfaces of constant radius are then surfaces of constant (positive) curvature which are everywhere perpendicular to the (radial) geodesics. Hence we obtain in invariant language:

There exists a family of surfaces orthogonal to the geodesics. Each of these surfaces is a surface of constant curvature. The segments of these geodesics contained between any two surfaces of the family are equal.

*Remark.* The case which has thus been obtained intuitively is not the general one in so far as the surfaces of the family could be of constant negative curvature or Euclidean (zero curvature).

* This condition not only limits the metric, but it necessitates that for every geodesic there exist a system of coordinates such that relative to this system the invariance under rotation around this geodesic is valid.

The four-dimensional case which interests us is entirely analogous. Furthermore there is no essential difference when the metric space is of index of inertia 1; only that one has to choose the radial directions as timelike and correspondingly the directions in the surfaces of the family as spacelike. The axes of the local light cones of all points lie on the radial lines.

### CHOICE OF COORDINATES

Instead of the four coordinates for which the spatial isotropy of the universe is most clearly apperent, we now choose different coordinates which are more convenient from the point of view of physical interpretation.

As timelike lines on which $x_1$, $x_2$, $x_3$ are constant and $x_4$ alone variable we choose the particle geodesics which in the central symmetric form are the straight lines through the center. Let $x_4$ further equal the metric distance from the center. In such coordinates the metric is of the form:

$$(2) \qquad \begin{cases} ds^2 = dx_4{}^2 - d\sigma^2 \\ d\sigma^2 = \gamma_{ik} dx_i dx_k \qquad (i, k = 1, 2, 3) \end{cases}$$

$d\sigma^2$ is the metric on one of the spherical hypersurfaces. The $\gamma_{ik}$ which belong to different hypersurfaces will then (because of the central symmetry) be the same form on all hypersurfaces except for a positive factor which depends on $x_4$ alone:

$$(2a) \qquad \gamma_{ik} = \underset{0}{\gamma_{ik}} G^2$$

where the $\underset{0}{\gamma}$ depend on $x_1$, $x_2$, $x_3$ only, and $G$ is a function of $x_4$ alone. We have then:

$$(2b) \qquad \underset{0}{d\sigma^2} = \underset{0}{\gamma_{ik}} dx_i dx_k \qquad (i, k = 1, 2, 3)$$

is a definite metric of constant curvature in three dimensions, the same for every $G$.

Such a metric is characterized by the equations:

$$\underset{0}{R_{iklm}} - B(\underset{0}{\gamma_{il}}\underset{0}{\gamma_{km}} - \underset{0}{\gamma_{im}}\underset{0}{\gamma_{kl}}) = 0 \tag{2c}$$

We can choose the coordinate system $(x_1, x_2, x_3)$ so that the line element becomes conformally Euclidean:

$$\underset{0}{d\sigma^2} = A^2(dx_1^2 + dx_2^2 + dx_3^2) \text{ i.e. } \underset{0}{\gamma_{ik}} = A^2\delta_{ik} \tag{2d}$$

where $A$ shall be a positive function of $r(r = x_1^2 + x_2^2 + x_3^2)$ alone. By substitution into the equations, we get for $A$ the two equations:

$$\left\{\begin{aligned} -\frac{1}{r}\left(\frac{A'}{Ar}\right)' + \left(\frac{A'}{Ar}\right)^2 &= 0 \\ -\frac{2A'}{Ar} - \left(\frac{A'}{A}\right)^2 - BA^2 &= 0 \end{aligned}\right. \tag{3}$$

The first equation is satisfied by:

$$A = \frac{c_1}{c_2 + c_3 r^2} \tag{3a}$$

where the constants are arbitrary for the time being. The second equation then yields:

$$B = 4\frac{c_2 c_3}{c_1^2} \tag{3b}$$

About the constants $c$ we get the following: If for $r = 0$, $A$ shall be positive, then $c_1$ and $c_2$ must have the same sign. Since a change of sign of all three constants does not change $A$, we can make $c_1$ and $c_2$ both positive. We can also make $c_2$ equal to 1. Furthermore, since a positive factor can always be incorporated into the $G^2$, we can also make $c_1$ equal to 1 without loss of generality. Hence we can set:

$$A = \frac{1}{1 + cr^2};\ B = 4c \tag{3c}$$

There are now three cases:

$c > 0$ (spherical space)
$c < 0$ (pseudospherical space)
$c = 0$ (Euclidean space)

By a similarity transformation of coordinates ($x_1' = ax_i$, where $a$ is constant), we can further get in the first case $c = \frac{1}{4}$, in the second case $c = -\frac{1}{4}$.

For the three cases we then have respectively:

$$\begin{cases} A = \dfrac{1}{1 + \dfrac{r^2}{4}};\ B = +1 \\ A = \dfrac{1}{1 - \dfrac{r^2}{4}};\ B = -1 \\ A = 1;\ B = 0 \end{cases} \tag{3d}$$

In the spherical case the "circumference" of the unit space ($G = 1$) is $\displaystyle\int_{-\infty}^{\infty} \frac{dr}{1 + \frac{r^2}{4}} = 2\pi$, the "radius" of the unit space is 1. In all three cases the function $G$ of time is a measure for the change with time of the distance of two points of matter (measured on a spatial section). In the spherical case, $G$ is the radius of space at the time $x_4$.

*Summary.* The hypothesis of *spatial* isotropy of our idealized universe leads to the metric:

(2) $$ds^2 = dx_4^2 - G^2A^2(dx_1^2 + dx_2^2 + dx_3^2)$$

where $G$ depends on $x_4$ alone, $A$ on $r$ $(= x_1^2 + x_2^2 + x_3^2)$ alone, where:

(3) $$A = \frac{1}{1 + \frac{z}{4}r^2}$$

and the different cases are characterized by $z = 1$, $z = -1$, and $z = 0$ respectively.

### THE FIELD EQUATIONS

We must now further satisfy the field equations of gravitation, that is to say the field equations without the "cosmologic member" which had been introduced previously *ad hoc:*

(4) $$(R_{ik} - \tfrac{1}{2}g_{ik}R) + \kappa T_{ik} = 0$$

By substitution of the expression for the metric, which was based on the assumption of spatial isotropy, we get after calculation:

$$R_{ik} - \frac{1}{2}g_{ik}R = \left(\frac{z}{G^2} + \frac{G'^2}{G^2} + 2\frac{G''}{G}\right)GA\delta_{ik}$$

$$(i, k = 1, 2, 3)$$

(4a) $$R_{44} - \frac{1}{2}g_{44}R = -3\left(\frac{z}{G^2} + \frac{G'^2}{G^2}\right)$$

$$R_{i4} - \tfrac{1}{2}g_{i4}R = 0 \qquad (i = 1, 2, 3)$$

Further we have for $T_{ik}$, the energy tensor of matter, for "dust":

(4b) $$T^{ik} = \rho\frac{dx_i}{ds}\frac{dx_k}{ds}$$

The geodesics, along which the matter moves, are the lines along which $x_4$ alone varies; on them $dx_4 = ds$. We have:

$$T^{44} = \rho \tag{4c}$$

the only component different from zero. By lowering of the indices we get as the only non-vanishing component of $T_{ik}$:

$$T_{44} = \rho \tag{4d}$$

Considering this, the field equations are:

$$\left\{\begin{aligned} &\frac{z}{G^2} + \frac{G'^2}{G^2} + 2\frac{G''}{G} = 0 \\ &\frac{z}{G^2} + \frac{G'^2}{G^2} - \frac{1}{3}\kappa\rho = 0 \end{aligned}\right. \tag{5}$$

$\frac{z}{G^2}$ is the curvature in the spatial section $x_4 = \text{const.}$ Since $G$ is in all cases a relative measure for the metric distance of two material particles as function of time, $\frac{G'}{G}$ expresses Hubble's expansion. $A$ drops out of the equations, as it has to if there shall be solutions of the equations of gravity of the required symmetrical type. By subtraction of both equations we get:

$$\frac{G''}{G} + \frac{1}{6}\kappa\rho = 0 \tag{5a}$$

Since $G$ and $\rho$ must be everywhere positive, $G''$ is everywhere negative for nonvanishing $\rho$. $G(x_4)$ can thus have no minimum nor a point of inflection; further there is no solution for which $G$ is constant.

## THE SPECIAL CASE OF VANISHING SPATIAL CURVATURE ($z = 0$)

The simplest special case for non-vanishing density $\rho$ is the case $z = 0$, where the sections $x_4$ = const are not curved. If we set $\frac{G'}{G} = h$, the field equations in this case are:

$$\text{(5b)} \qquad \left\{ \begin{aligned} 2h' + 3h^2 &= 0 \\ 3h^2 &= \kappa\rho \end{aligned} \right.$$

The relation between Hubble's expansion $h$ and the average density $\rho$, which is given in the second equation, is comparable to some extent with experience, at least as far as the order of magnitude is concerned. The expansion is given as 432 km/sec for the distance of $10^6$ parsec. If we express this in the system of measures used by us (cm—as unit length; unit of time—that of motion of light of one cm) we get:

$$h = \frac{432 \cdot 10^5}{3.25 \cdot 10^6 \cdot 365 \cdot 24 \cdot 60 \cdot 60} \cdot \left(\frac{1}{3 \cdot 10^{10}}\right)^2 = 4.71 \cdot 10^{-28}.$$

Since further (see 105a) $\kappa = 1.86\ 10^{-27}$, the second equation of (5b) yields:

$$\rho = \frac{3h^2}{\kappa} = 3.5 \cdot 10^{-28} \text{ g./cm.}^3$$

This value corresponds, according to the order of magnitude, somewhat with the estimates given by astronomers (on the basis of the masses and parallaxes of visible stars and systems of stars). I quote here as example G. C. McVittie (Proceedings of the Physical Society of London, vol. 51, 1939, p. 537): "The average density is certainly not greater than $10^{-27}$ g./cm.$^3$ and is more probably of the order $10^{-29}$ g./cm.$^3$"

Owing to the great difficulty of determining this magnitude I consider this for the time being a satisfactory correspondence. Since the quantity $h$ is determined with greater accuracy than $\rho$, it is probably not an exaggeration to assert that the determination of the structure of observable space is tied up with the more precise determination of $\rho$. Because, due to the second equation of (5), the space curvature is given in the general case as:

$$zG^{-2} = \tfrac{1}{3}\kappa\rho - h^2. \tag{5c}$$

Hence, if the right side of the equation is positive, the space is of positive constant curvature and therefor finite; its magnitude can be determined with the same accuracy as this difference. If the right side is negative, the space is infinite. At present $\rho$ is not sufficiently determined to enable us to deduce from this relation a non-vanishing mean curvature of space (the section $x_4$ = const).

In case we neglect spatial curvature, the first equation of (5c) becomes, after suitable choice of the initial point of $x_4$:

$$h = \frac{2}{3} \cdot \frac{1}{x_4} \tag{6}$$

This equation has a singularity for $x_4 = 0$, so that such a space has either a negative expansion and the time is limited from above by the value $x_4 = 0$, or it has a positive expansion and begins to exist for $x_4 = 0$. The latter case corresponds to what we find realized in nature.

From the measured value of $h$ we get for the time of existence of the world up to now $1.5 \cdot 10^9$ years. This age is about the same as that which one has obtained from the disintegration of uranium for the firm crust of the earth. This is a paradoxical result, which for more than one reason has aroused doubts as to the validity of the theory.

The question arises: Can the present difficulty, which arose under the assumption of a practically negligible spatial curvature, be eliminated by the introduction of a suitable spatial curvature? Here the first equation of (5), which determines the time-dependence of $G$, will be of use.

### SOLUTION OF THE EQUATIONS IN THE CASE OF NON-VANISHING SPATIAL CURVATURE

If one considers a spatial curvature of the spatial section ($x_4$ = const), one has the equations:

$$(5)\qquad \begin{aligned} zG^{-2} + \left(2\frac{G''}{G} + \left(\frac{G'}{G}\right)^2\right) &= 0 \\ zG^{-2} + \left(\frac{G'}{G}\right)^2 - \frac{1}{3}\kappa\rho &= 0 \end{aligned}$$

The curvature is positive for $z = +1$, negative for $z = -1$. The first of these equations is integrable. We first write it it in the form:

$$(5d)\qquad z + 2GG'' + G'^2 = 0.$$

If we consider $x_4$ $(= t)$ as a function of $G$, we have:

$$G' = \frac{1}{t'},\ G'' = \left(\frac{1}{t'}\right)' \frac{1}{t'}.$$

If we write $u(G)$ for $\frac{1}{t'}$, we get:

$$(5e)\qquad z + 2Guu' + u^2 = 0$$

or

$$(5f)\qquad z + (Gu^2)' = 0.$$

From this we get by simple integration:

$$zG + Gu^2 = G_0 \tag{5g}$$

or, since we set $u = \frac{1}{\frac{dt}{dG}} = \frac{dG}{dt}$:

$$\left(\frac{dG}{dt}\right)^2 = \frac{G_0 - zG}{G} \tag{5h}$$

where $G_0$ is a constant. This constant cannot be negative, as we see if we differentiate (5h) and consider that $G''$ is negative because of (5a).

(a) Space with positive curvature

$G$ remains in the interval $0 \leq G \leq G_0$. $G$ is given quantitatively by a sketch like the following:

(1)

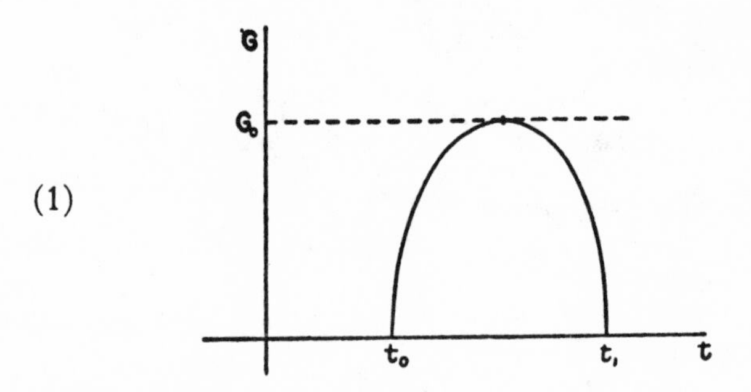


The radius $G$ rises from 0 to $G_0$ and then again drops continuously to 0. The spatial section is finite (spherical)

$$\tfrac{1}{3}\kappa\rho - h^2 > 0 \tag{5c}$$

(b) Space with negative curvature

$$\left(\frac{dG}{dt}\right)^2 = \frac{G_0 + G}{G}.$$

$G$ increases with $t$ from $G = 0$ to $G = +\infty$ (or goes from $G = \infty$ to $G = 0$). Hence $\frac{dG}{dt}$ decreases monotonically from $+\infty$ to 1 as illustrated by the sketch:

(2)

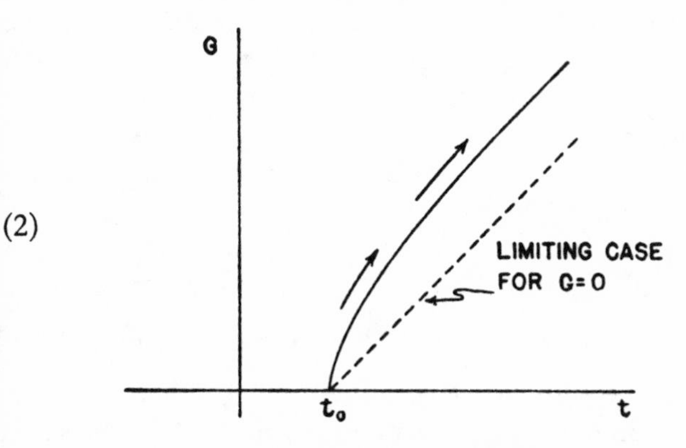


This is then a case of continued expansion with no contraction. The spatial section is infinite, and we have:

(5c) $$\tfrac{1}{3}\kappa\rho - h^2 < 0.$$

The case of plane spatial section, which was treated in the previous section, lies between these two cases, according to the equation:

(5h) $$\left(\frac{dG}{dt}\right)^2 = \frac{G_0}{G}.$$

*Remark.* The case of negative curvature contains as a limiting case that of vanishing $\rho$. For this case $\left(\frac{dG}{dt}\right)^2 = 1$ (see sketch 2). This is the Euclidean case; since the calculations show that the curvature tensor vanishes.

The case of negative curvature with non-vanishing $\rho$ approaches this limiting case more and more closely, so that with increasing time the structure of space will be less and less determined by the matter contained in it.

From this investigation of the case of non-vanishing curvature results the following. For every state of non-vanishing ("spatial") curvature, there exists, as in the case of vanishing curvature, an initial state in which $G = 0$ where the expansion starts. Hence this is a section at which the density is infinite and the field is singular. The introduction of such a new singularity seems problematical in itself.*

It appears, further, that the influence of the introduction of a spatial curvature on the time interval between the start of the expansion and the drop to a fixed value $h = \frac{G'}{G}$ is negligible in its order of magnitude. This time interval may be obtained by elementary calculations from (5$h$), which we shall omit here. We restrict ourselves to the consideration of an expanding space with vanishing $\rho$. This, as mentioned before, is a special case of negative spatial curvature. The second equation of (5) yields (considering the reversal of sign of the first member):

$$G' = 1.$$

Hence (for suitable initial point for $x_4$)

$$G = x_4$$

$$h = \frac{G'}{G} = \frac{1}{x_4} \quad . \ . \ . \tag{6a}$$

Hence this extreme case yields for the duration of the expansion the same result as the case of vanishing spatial

* However, the following should be noted: The present relativistic theory of gravitation is based on a separation of the concepts of "gravitational field" and of "matter." It may be plausible that the theory is for this reason inadequate for very high density of matter. It may well be the case that for a unified theory there would arise no singularity.

curvature (see Eq. 6) except for a factor of order of magnitude 1.

The doubts mentioned in connection with Eq. (6), namely that this would give such a remarkably short duration for the development of the stars and systems of stars which are observable at present, cannot therefore be removed by the introduction of a spatial curvature.

### EXTENSION OF THE ABOVE CONSIDERATIONS BY GENERALIZATION OF THE EQUATION WITH RESPECT TO PONDERABLE MATTER

For all the solutions reached up to now, there exists a state of the system at which the metric becomes singular ($G = 0$) and the density $\rho$ becomes infinite. The following question arises: Could not the rise of such singularities be due to the fact that we introduced matter as a kind of dust which does not resist condensation? Did we not neglect, without justification, the influence of the random motion of the single stars?

One could, for example, replace dust whose particles are at rest relative to each other, by one whose particles are in random motion relative to each other like the molecules of a gas. Such matter would offer a resistance to adiabatic condensation which increases with that condensation. Will not this be able to prevent the rise of infinite condensation? We shall show below that such a modification in the description of matter can change nothing of the main character of the above solutions.

### "PARTICLE-GAS" TREATED ACCORDING TO SPECIAL RELATIVITY

We consider a swarm of particles of mass $m$ in parallel motion. By a proper transformation this swarm can be considered at rest. The spatial density of the particles, $\sigma$, is then invariant in the Lorentz sense. Related to an arbitrary Lorentz system

$$T^{uv} = m\sigma \frac{dx^u}{ds}\frac{dx^v}{ds} \tag{7}$$

has invariant meaning (energy tensor of the swarm). If there exist many such swarms we get, by summation, for all of them:

$$T^{uv} = m \sum_p \sigma_p \left(\frac{dx^u}{ds}\right)_p \left(\frac{dx^v}{ds}\right)_p \tag{7a}$$

In relation to this form we can choose the time axis of the Lorentz system so that: $T^{14} = T^{24} = T^{34} = 0$. Further, we can obtain by spatial rotation of the system: $T^{12} = T^{23} = T^{31} = 0$. Let, further, the particle gas be isotropic. This means that $T^{11} = T^{22} = T^{33} = p$. This is an invariant as well as $T^{44} = u$. The invariant:

$$J = T^{uv}g_{uv} = T^{44} - (T^{11} + T^{22} + T^{33}) = u - 3p \tag{7b}$$

is thus expressed in terms of $u$ and $p$.

It follows from the expression for $T^{uv}$ that $T^{11}$, $T^{22}$, $T^{33}$ and $T^{44}$ are all positive; hence the same is true for $T_{11}$, $T_{22}$, $T_{33}$, $T_{44}$.

The equations of gravity are now:

$$\begin{aligned} 1 + 2GG'' + G^2 + \kappa T_{11} &= 0 \\ -3G^{-2}(1 + G'^2) + \kappa T_{44} &= 0. \end{aligned} \tag{8}$$

From the first it follows that here too (since $T_{11} > 0$) $G''$ is always negative where the member $T_{11}$ for given $G$ and $G'$ can only decrease the value of $G''$.

From this we see that the consideration of a random relative motion of the mass points does not change our results fundamentally.

### SUMMARY AND OTHER REMARKS

(1) The introduction of the "cosmologic member" into the equations of gravity, though possible from the point of view of relativity, is to be rejected from the point of view of logical economy. As Friedman was the first to show one can reconcile an everywhere finite density of matter with the original form of the equations of gravity if one admits the time variability of the metric distance of two mass points.*

(2) The demand for *spatial* isotropy of the universe alone leads to Friedman's form. It is therefore undoubtedly the general form, which fits the cosmologic problem.

(3) Neglecting the influence of spatial curvature, one obtains a relation between the mean density and Hubble's expansion which, as to order of magnitude, is confirmed empirically.

One further obtains, for the time from the start of the expansion up to the present, a value of the order of magnitude of $10^9$ years. The brevity of this time does not concur with the theories on the developments of fixed stars.

(4) The latter result is not changed by the introduction of spatial curvature; nor is it changed by the consideration of the random motion of stars and systems of stars with respect to each other.

* If Hubble's expansion had been discovered at the time of the creation of the general theory of relativity, the cosmologic member would never have been introduced. It seems now so much less justified to introduce such a member into the field equations, since its introduction loses its sole original justification,—that of leading to a natural solution of the cosmologic problem.

(5) Some try to explain Hubble's shift of spectral lines by means other than the Doppler effect. There is, however, no support for such a conception in the known physical facts. According to such a hypothesis it would be possible to connect two stars, $S_1$ and $S_2$, by a rigid rod. Monochromatic light which is sent from $S_1$ to $S_2$ and reflected back to $S_1$ could arrive with a different frequency (measured by a clock on $S_1$) if the number of wave lengths of light along the rod should change with time on the way. This would mean that the locally measured velocity of light would depend on time, which would contradict even the special theory of relativity. Further it should be noted that a light signal going to and fro between $S_1$ and $S_2$ would constitute a "clock" which would not be in a constant relation with a clock (e.g. an atomistic clock) in $S_1$. This would mean that there would exist no metric in the sense of relativity. This not only involves the loss of comprehension of all those relations which relativity has yielded, but it also fails to concur with the fact that certain atomistic forms are not related by "similarity" but by "congruence" (the existence of sharp spectral lines, volumes of atoms, etc.).

The above considerations are, however, based on wave theory, and it may be that some proponents of the above hypothesis imagine that the process of the expansion of light is altogether not according to wave theory, but rather in a manner analogous to the Compton effect. The assumption of such a process without scattering constitutes a hypothesis which is not justified from the point of view of our present knowledge. It also fails to give a reason for the independence of the relative shift of frequence from the original frequency. Hence one cannot but consider Hubble's discovery as an expansion of the system of stars.

(6) The doubts about the assumption of a "beginning of the world" (start of the expansion) only about $10^9$ years ago have roots of both an empirical and a theoretical nature. The astronomers tend to consider the stars of different spectral types as age classes of a uniform development, which process would need much longer than $10^9$ years. Such a theory therefore actually contradicts the demonstrated consequences of the relativistic equations. It seems to me, however, that the "theory of evolution" of the stars rests on weaker foundations than the field equations.

The theoretical doubts are based on the fact that for the time of the beginning of the expansion the metric becomes singular and the density, $\rho$, becomes infinite. In this connection the following should be noted: The present theory of relativity is based on a division of physical reality into a metric field (gravitation) on the one hand, and into an electromagnetic field and matter on the other hand. In reality space will probably be of a uniform character and the present theory be valid only as a limiting case. For large densities of field and of matter, the field equations and even the field variables which enter into them will have no real significance. One may not therefore assume the validity of the equations for very high density of field and of matter, and one may not conclude that the "beginning of the expansion" must mean a singularity in the mathematical sense. All we have to realize is that the equations may not be continued over such regions.

This consideration does, however, not alter the fact that the "beginning of the world" really constitutes a beginning, from the point of view of the development of the now existing stars and systems of stars, at which those stars and systems of stars did not yet exist as individual entities.

(7) There are, however, some empirical arguments in *favor* of a dynamic concept of space as required by the theory. Why does there still exist uranium, despite its comparatively rapid decomposition, and despite the fact that no possibility for the creation of uranium is recognizable? Why is space not so filled with radiation as to make the nocturnal sky look like a glowing surface? This is an old question which so far has found no satisfactory answer from the point of view of a stationary world. But it would lead too far to go into questions of this type.

(8) For the reasons given it seems that we have to take the idea of an expanding universe seriously, in spite of the short "lifetime." If one does so, the main question becomes whether space has positive or negative spatial curvature. To this we add the following remark.

From the empirical point of view the decision boils down to the question whether the expression $\frac{1}{3}\kappa\rho - h^2$ is positive (spherical case) or negative (pseudospherical case). This seems to me to be the most important question. An empirical decision does not seem impossible at the present state of astronomy. Since $h$ (Hubble's expansion) is comparatively well known, everything depends on determining $\rho$ with the highest possible accuracy.

It is imaginable that the proof would be given that the world is spherical (it is hardly imaginable that one could prove it to be pseudospherical). This depends on the fact that one can always give a lower bound for $\rho$ but not an upper bound. This is the case because we can hardly form an opinion on how large a fraction of $\rho$ is given by astronomically unobservable (not radiating) masses. This I wish to discuss in somewhat greater detail.

One can give a lower bound for $\rho$ ($\rho_s$) by taking into consideration only the masses of radiating stars. If it should appear that $\rho_s > \frac{3h^2}{\kappa}$ then one would have decided in favor of spherical space. If it appears that $\rho_s < \frac{3h^2}{\kappa}$ one has to try to determine the share of non-radiating masses $\rho_d$. We want to show that one can also find a lower bound for $\frac{\rho_d}{\rho_s}$.

We consider an astronomical object which contains many single stars and which can be considered with sufficient accuracy to be a stationary system, e.g. a globular cluster (of known parallax). From the velocities which are observable spectroscopically one can determine the field of gravitation (under plausible assumptions) and thereby the masses which generate this field. The masses which are so computed one can compare with those of the visible stars of the cluster, and so find at least a rough approximation for how far the masses which generate the field exceed those of the visible stars of the cluster. One obtains thus an estimate for $\frac{\rho_d}{\rho_s}$ for the particular cluster.

Since the non-radiating stars will on the average be smaller than the radiating ones, they will tend on the average to greater velocities than the larger stars due to their interaction with the stars of the cluster. Hence they will "evaporate" more quickly from the cluster than the larger stars. It may therefore be expected that the relative frequency of the smaller heavenly bodies inside the cluster will be smaller than that outside of it. One can therefore obtain in $\left(\frac{\rho_d}{\rho_s}\right)_k$ (relation of densities in the above cluster) a lower bound for the ratio $\frac{\rho_d}{\rho_s}$ in the whole space. One therefore obtains as a lower bound for the entire average density of mass in space:

$$\rho_s \left[ 1 + \left( \frac{\rho_d}{\rho_s} \right)_k \right]$$

If this quantity is greater than $\frac{3h^2}{\kappa}$ one may conclude that space is of a spherical character. On the other hand I cannot think of any reasonably reliable determination of an upper bound for $\rho$.

(9) Last and not least: The age of the universe, in the sense used here, must certainly exceed that of the firm crust of the earth as found from the radioactive minerals. Since determination of age by these minerals is reliable in every respect, the cosmologic theory here presented would be disproved if it were found to contradict any such results. In this case I see no reasonable solution.

# APPENDIX II

# RELATIVISTIC THEORY OF THE NON-SYMMETRIC FIELD

BEFORE starting with the subject proper I am first going to discuss the "strength" of systems of field equations in general. This discussion is of intrinsic interest quite apart from the particular theory presented here. For a deeper understanding of our problem, however, it is almost indispensable.

## ON THE "COMPATIBILITY" AND THE "STRENGTH" OF SYSTEMS OF FIELD EQUATIONS

Given certain field variables and a system of field equations for them, the latter will not in general determine the field completely. There still remain certain free data for a solution of the field equations. The smaller the number of free data consistent with the system of field equations, the "stronger" is the system. It is clear that in the absence of any other viewpoint from which to select the equations, one will prefer a "stronger" system to a less strong one. It is our aim to find a measure for this strength of systems of equations. It will turn out that such a measure can be defined which will even enable us to compare with each other the strengths of systems whose field variables differ with respect to number and kind.

We shall present the concepts and methods involved here in examples of increasing complexity, restricting ourselves to four-dimensional fields, and in the course of these examples we shall successively introduce the relevant concepts.

# APPENDIX II

First example: *The scalar wave equation**

$$\phi_{,11} + \phi_{,22} + \phi_{,33} - \phi_{,44} = 0.$$

Here the system consists of only *one* differential equation for *one* field variable. We assume $\phi$ to be expanded in a Taylor series in the neighborhood of a point $P$ (which presupposes the analytic character of $\phi$). The totality of its coefficients describes then the function completely. The number of $n$th order coefficients (that is, the $n$th order derivatives of $\phi$ at the point $P$) is equal to $\frac{4 \cdot 5 \ldots (n+3)}{1 \cdot 2 \ldots n}$ $\left(\text{abbreviated } \binom{4}{n}\right)$, and all these coefficients could be freely chosen if the differential equation did not imply certain relations between them. Since the equation is of second order, these relations are found by $(n-2)$ fold differentiation of the equation. We thus obtain for the $n$th order coefficients $\binom{4}{n-2}$ conditions. The number of $n$th order coefficients remaining free is therefore

$$z = \binom{4}{n} - \binom{4}{n-2}. \tag{1}$$

This number is positive for any $n$. Hence, if the free coefficients for all orders smaller than $n$ have been fixed, the conditions for the coefficients of order $n$ can always be satisfied without changing the coefficients already chosen.

Analogous reasoning can be applied to systems consisting of several equations. If the number of free $n$th order coefficients does not become smaller than zero, we call the system of equations *absolutely compatible*. We shall

* In the following the comma will always denote partial differentiation; thus, for example, $\phi_{,i} = \frac{\partial \phi}{\partial x^i}$, $\phi_{,11} = \frac{\partial^2 \phi}{\partial x^1 \partial x^1}$ etc.

restrict ourselves to such systems of equations. All systems known to me which are used in physics are of this kind.

Let us now rewrite equation (1). We have

$$\binom{4}{n-2} = \binom{4}{n}\frac{(n-1)n}{(n+2)\,(n+3)} = \binom{4}{n}\left(1 - \frac{z_1}{n} + \frac{z_2}{n^2} + \ldots\right)$$

where $z_1 = +6$.

If we restrict ourselves to large values of $n$, we may neglect the terms $\frac{z_2}{n^2}$ etc. in the parenthesis, and we obtain for (1) *asymptotically*

$$z \sim \binom{4}{n}\frac{z_1}{n} = \binom{4}{n}\frac{6}{n}. \qquad \text{(1a)}$$

We call $z_1$ the "coefficient of freedom," which in our case has the value 6. The larger this coefficient, the weaker is the corresponding system of equations.

Second example: *Maxwell's equations for empty space*

$$\phi^{is}{}_{,s} = 0; \qquad \phi_{ik,l} + \phi_{kl,i} + \phi_{li,k} = 0.$$

$\phi^{ik}$ results from the antisymmetric tensor $\phi_{ik}$ by raising the covariant indices with the help of

$$\eta^{ik} = \begin{pmatrix} -1 & & & \\ & -1 & & \\ & & -1 & \\ & & & +1 \end{pmatrix}.$$

These are $4 + 4$ field equations for six field variables. Among these eight equations, there exist two identities. If the left-hand sides of the field equations are denoted by $G^i$ and $H_{ikl}$ respectively, the identities have the form

$$G^i{}_{,i} \equiv 0; \qquad H_{ikl,m} - H_{klm,i} + H_{lmi,k} - H_{mik,l} = 0.$$

In this case we reason as follows.

# APPENDIX II

The Taylor expansion of the six field components furnishes

$$6\binom{4}{n}$$

coefficients of the $n$th order. The conditions that these $n$th order coefficients must satisfy are obtained by $(n-1)$fold differentiation of the eight field equations of the first order. The number of these conditions is therefore

$$8\binom{4}{n-1}.$$

These conditions, however, are not independent of each other, since there exist among the eight equations two identities of second order. They yield upon $(n-2)$fold differentiation

$$2\binom{4}{n-2}$$

algebraic identities among the conditions obtained from the field equations. The number of free coefficients of $n$th order is therefore

$$z = 6\binom{4}{n} - \left[8\binom{4}{n-1} - 2\binom{4}{n-2}\right].$$

$z$ is positive for all $n$. The system of equations is thus "absolutely compatible." If we extract the factor $\binom{4}{n}$ on the right-hand side and expand as above for large $n$, we obtain asymptotically

$$z = \binom{4}{n}\left[6 - 8\frac{n}{n+3} + 2\frac{(n-1)n}{(n+2)(n+3)}\right]$$

$$\sim \binom{4}{n}\left[6 - 8\left(1 - \frac{3}{n}\right) + 2\left(1 - \frac{6}{n}\right)\right]$$

$$\sim \binom{4}{n}\left[0 + \frac{12}{n}\right].$$

Here, then, $z_1 = 12$. This shows that—and to what extent—this system of equations determines the field less strongly than in the case of the scalar wave equation ($z_1 = 6$). The circumstance that in both cases the constant term in the parenthesis vanishes expresses the fact that the system in question does not leave free any function of four variables.

Third example: *The gravitational equations for empty space.* We write them in the form

$$R_{ik} = 0; \qquad g_{ik,l} - g_{sk}\,\Gamma^s_{il} - g_{is}\,\Gamma^s_{lk} = 0.$$

The $R_{ik}$ contain only the $\Gamma$ and are of first order with respect to them. We treat here the $g$ and $\Gamma$ as independent field variables. The second equation shows that it is convenient to treat the $\Gamma$ as quantities of the first order of differentiation, which means that in the Taylor expansion

$$\Gamma = \underset{0}{\Gamma} + \underset{1}{\Gamma_s}\, x^s + \underset{2}{\Gamma_{st}}\, x^s\, x^t + \ldots$$

we consider $\underset{0}{\Gamma}$ to be of the first order, $\underset{1}{\Gamma_s}$ of the second order, and so on. Accordingly, the $R_{ik}$ must be considered as of second order. Between these equations, there exist the four Bianchi identities which, as a consequence of the convention adopted, are to be considered as of third order.

In a generally covariant system of equations a new circumstance appears which is essential for a correct enumeration of the free coefficients: fields that result from one another by mere coordinate transformations should be considered only as different representations of one and the same field. Correspondingly, only part of the

$$10\binom{4}{n}$$

$n$th order coefficients of the $g_{ik}$ serves to characterize essentially different fields. Therefore, the number of expansion coefficients that actually determine the field is reduced by a certain amount which we must now compute.

In the transformation law for the $g_{ik}$,

$$g_{ik}{}^* = \frac{\partial x^a}{\partial x^{i^*}} \frac{\partial x^b}{\partial x^{k^*}} g_{ab}$$

$g_{ab}$ and $g_{ik}{}^*$ represent in fact the same field. If this equation is differentiated $n$ times with respect to the $x^*$, one notices that all $(n+1)$st derivatives of the four functions $x$ with respect to the $x^*$ enter into the $n$th order coefficients of the $g^*$-expansion; i.e., there appear $4\binom{4}{n+1}$ numbers that have no part in the characterization of the field. In any general-relativistic theory one must therefore subtract $4\binom{4}{n+1}$ from the total number of $n$th order coefficients so as to take account of the general covariance of the theory. The enumeration of the free coefficients of $n$th order leads thus to the following result.

The ten $g_{ik}$ (quantities of zero order of differentiation) and the forty $\Gamma^l_{ik}$ (quantities of first order of differentiation) yield in view of the correction just derived

$$10\binom{4}{n} + 40\binom{4}{n-1} - 4\binom{4}{n+1}$$

relevant coefficients of $n$th order. The field equations (10 of the second and 40 of the first order) furnish for them

$$N = 10\binom{4}{n-2} + 40\binom{4}{n-1}$$

conditions. From this number, however, we must subtract

the number of the identities between these $N$ conditions, viz.

$$4\binom{4}{n-3}$$

which result from the Bianchi identities (of the third order). Hence we find here

$$z = \left[10\binom{4}{n} + 40\binom{4}{n-1} - 4\binom{4}{n+1}\right] - \left[10\binom{4}{n-2} + 40\binom{4}{n-1}\right] + 4\binom{4}{n-3}.$$

Extracting again the factor $\binom{4}{n}$ we obtain asymptotically for large $n$

$$z \sim \binom{4}{n}\left[0 + \frac{12}{n}\right]. \quad \text{Thus } z_1 = 12.$$

Here, too, $z$ is positive for all $n$ so that the system is absolutely compatible in the sense of the definition given above. It is surprising that the gravitational equations for empty space determine their field just as strongly as do Maxwell's equations in the case of the electromagnetic field.

## RELATIVISTIC FIELD THEORY

### *General remarks*

It is the essential achievement of the general theory of relativity that it has freed physics from the necessity of introducing the "inertial system" (or inertial systems). This concept is unsatisfactory for the following reason: without any deeper foundation it singles out certain co-ordinate systems among all conceivable ones. It is then assumed that the laws of physics hold *only* for such inertial systems (e.g. the law of inertia and the law of the constancy of the velocity of light). Thereby, space as such is assigned a role in the system of physics that distinguishes it from all other elements of physical description. It plays a determining role in all processes, without in its turn being influenced by them. Though such a theory is logically possible, it is on the other hand rather unsatisfactory. Newton had been fully aware of this deficiency, but he had also clearly understood that no other path was open to physics in his time. Among the later physicists it was above all Ernst Mach who focussed attention on this point.

What innovations in the post-Newtonian development of the foundations of physics have made it possible to overcome the inertial system? First of all, it was the introduction of the field concept by, and subsequent to, the theory of electromagnetism of Faraday and Maxwell, or to be more precise, the introduction of the field as an independent, not further reducible fundamental concept. As far as we are able to judge at present, the general theory of relativity can be conceived only as a field theory. It could not have developed if one had held on to the view that the real world consists of material points which move under the influence of forces acting between them. Had one tried to explain to Newton the equality of inertial and gravitational mass from the equivalence principle, he would necessarily have had to reply with the following objection: it is indeed true that relative to an accelerated coordinate system bodies experience the same accelerations as they do relative to a gravitating celestial body close to its surface. But where are, in the former case, the masses that produce the accelerations? It is clear that the theory of relativity presupposes the independence of the field concept.

The mathematical knowledge that has made it possible

to establish the general theory of relativity we owe to the geometrical investigations of Gauss and Riemann. The former has, in his theory of surfaces, investigated the metric properties of a surface imbedded in three-dimensional Euclidean space, and he has shown that these properties can be described by concepts that refer only to the surface itself and not to its relation to the space in which it is imbedded. Since, in general, there exists no preferred coordinate system on a surface, this investigation led for the first time to expressing the relevant quantities in general coordinates. Riemann has extended this two-dimensional theory of surfaces to spaces of an arbitrary number of dimensions (spaces with Riemannian metric, which is characterized by a symmetric tensor field of second rank). In this admirable investigation he found the general expression for the curvature in higher-dimensional metric spaces.

The development just sketched of the mathematical theories essential for the setting up of general relativity had the result that at first Riemannian metric was considered the fundamental concept on which the general theory of relativity and thus the avoidance of the inertial system were based. Later, however, Levi-Cività rightly pointed out that the element of the theory that makes it possible to avoid the inertial system is rather the infinitesimal displacement field $\Gamma^l_{ik}$. The metric or the symmetric tensor field $g_{ik}$ which defines it is only indirectly connected with the avoidance of the inertial system in so far as it determines a displacement field. The following consideration will make this clear.

The transition from one inertial system to another is determined by a *linear* transformation (of a particular kind).

# APPENDIX II

If at two arbitrarily distant points $P_1$ and $P_2$ there are two vectors $\underset{1}{A^i}$ and $\underset{2}{A^i}$ respectively whose corresponding components are equal to each other $(\underset{1}{A^i} = \underset{2}{A^i})$, this relation is conserved in a permissible transformation. If in the transformation formula

$$A^{i^*} = \frac{\partial x^{i^*}}{\partial x^a} A^a$$

the coefficients $\frac{\partial x^{i^*}}{\partial x^a}$ are independent of the $x^a$, the transformation formula for the vector components is independent of position. Equality of the components of two vectors at different points $P_1$ and $P_2$ is thus an invariant relation if we restrict ourselves to inertial systems. If, however, one abandons the concept of the inertial system and thus admits arbitrary continuous transformations of the coordinates so that the $\frac{\partial x^{i^*}}{\partial x^a}$ depend on the $x^a$, the equality of the components of two vectors attached to two different points in space loses its invariant meaning and thus vectors at different points can no longer be directly compared. It is due to this fact that in a general relativistic theory one can no longer form new tensors from a given tensor by simple differentiation and that in such a theory there are altogether much fewer invariant formations. This paucity is remedied by the introduction of the infinitesimal displacement field. It replaces the inertial system inasmuch as it makes it possible to compare vectors at infinitesimally close points. Starting from this concept we shall present in the sequel the relativistic field theory, carefully dispensing with anything that is not necessary to our purpose.

# APPENDIX II

## *The infinitesimal displacement field* $\Gamma$

To a contravariant vector $A^i$ at a point $P$ (coordinates $x^t$) we correlate a vector $A^i + \delta A^i$ at the infinitesimally close point $(x^t + dx^t)$ by the bilinear expression

$$\delta A^i = -\Gamma^i_{st} A^s dx^t \tag{2}$$

where the $\Gamma$ are functions of $x$. On the other hand, if $A$ is a vector field, the components of $(A^i)$ at the point $(x^t + dx^t)$ are equal to $A^i + dA^i$ where*

$$dA^i = A^i{}_{,t}\, dx^t.$$

The difference of these two vectors at the neighboring point $x^t + dx^t$ is then itself a vector

$$(A^i{}_{,t} + A^s \Gamma^i_{st})\, dx^t \equiv A^i_t\, dx^t$$

connecting the components of the vector field at two infinitesimally close points. The displacement field replaces the inertial system inasmuch as it effects this connection formerly furnished by the inertial system. The expression within the parenthesis, $A^i_t$ for short, is a tensor.

The tensor character of $A^i_t$ determines the transformation law for the $\Gamma$. We have first

$$A^i_k{}^* = \frac{\partial x^{i^*}}{\partial x^i} \frac{\partial x^k}{\partial x^{k^*}} A^i_k\,.$$

Using the same index in both coordinate systems is not meant to imply that it refers to corresponding components, i.e. $i$ in $x$ and in $x^*$ run *independently* from 1 to 4. After

*As before "$_{,t}$" denotes ordinary differentiation $\frac{\partial}{\partial x^t}$.

some practice this notation makes the equations considerably more transparent. We now replace

$$A^{i}_{k}{}^{*} \text{ by } A^{i^*}{}_{,k^*} + A^{s^*}\Gamma^{i}_{sk}{}^{*}$$

$$A^{i}_{k} \text{ by } A^{i}{}_{,k} + A^{s}\Gamma^{i}_{sk}$$

and again $A^{i^*}$ by $\dfrac{\partial x^{i^*}}{\partial x^{i}}A^{i}$, $\dfrac{\partial}{\partial x^{k^*}}$ by $\dfrac{\partial x^{k}}{\partial x^{k^*}}\cdot\dfrac{\partial}{\partial x^{k}}$.

This leads to an equation which, apart from the $\Gamma^*$, contains only field quantities of the original system and their derivatives with respect to the $x$ of the original system. Solving this equation for the $\Gamma^*$ one obtains the desired transformation formula

$$\Gamma^{i}_{kl}{}^{*} = \frac{\partial x^{i^*}}{\partial x^{i}}\frac{\partial x^{k}}{\partial x^{k^*}}\frac{\partial x^{l}}{\partial x^{l^*}}\Gamma^{i}_{kl} - \frac{\partial^2 x^{i^*}}{\partial x^{s}\partial x^{t}}\frac{\partial x^{s}}{\partial x^{k^*}}\frac{\partial x^{t}}{\partial x^{l^*}} \tag{3}$$

whose second term (on the right-hand side) can be somewhat simplified:

$$-\frac{\partial^2 x^{i^*}}{\partial x^{s}\partial x^{t}}\frac{\partial x^{s}}{\partial x^{k^*}}\frac{\partial x^{t}}{\partial x^{l^*}}$$

$$= -\frac{\partial}{\partial x^{l^*}}\left(\frac{\partial x^{i^*}}{\partial x^{s}}\right)\frac{\partial x^{s}}{\partial x^{k^*}} = -\frac{\partial}{\partial x^{l^*}}\left(\frac{\partial x^{i^*}}{\partial x^{k^*}}\right) + \frac{\partial x^{i^*}}{\partial x^{s}}\frac{\partial^2 x^{s}}{\partial x^{k^*}\partial x^{l^*}}$$

$$= \frac{\partial x^{i^*}}{\partial x^{s}}\frac{\partial^2 x^{s}}{\partial x^{k^*}\partial x^{l^*}} \tag{3a}$$

We call such a quantity a *pseudo tensor*. Under linear transformations it transforms as a tensor, whereas for non-linear transformations a term is added which does not contain the expression to be transformed, but only depends on the transformation coefficients.

# APPENDIX II

*Remarks on the displacement field.*

1. The quantity $\tilde{\Gamma}^i_{kl}$ $(\equiv \Gamma^i_{lk})$ which is obtained by a transposition of the lower indices also transforms according to (3) and is therefore likewise a displacement field.

2. By symmetrizing or anti-symmetrizing equation (3) with respect to the lower indices $k^*$, $l^*$ one obtains the two equations

$$\Gamma^i_{\underline{kl}}{}^* \left(= \tfrac{1}{2}(\Gamma^i_{kl}{}^* + \Gamma^i_{lk}{}^*)\right) = \frac{\partial x^{i^*}}{\partial x^i} \frac{\partial x^k}{\partial x^{k^*}} \frac{\partial x^l}{\partial x^{l^*}} \Gamma^i_{\underline{kl}} - \frac{\partial^2 x^{i^*}}{\partial x^s \partial x^t} \frac{\partial x^s}{\partial x^{k^*}} \frac{\partial x^t}{\partial x^{l^*}}$$

$$\Gamma^i_{\underset{\vee}{kl}}{}^* \left(= \tfrac{1}{2}(\Gamma^i_{kl}{}^* - \Gamma^i_{lk}{}^*)\right) = \frac{\partial x^{i^*}}{\partial x^i} \frac{\partial x^k}{\partial x^{k^*}} \frac{\partial x^l}{\partial x^{l^*}} \Gamma^i_{\underset{\vee}{kl}} .$$

Hence the two (symmetric and anti-symmetric) constituents of $\Gamma^i_{kl}$ transform independently of each other, i.e. without mixing. Thus they appear from the point of view of the transformation law as independent quantities. The second of the equations shows that $\Gamma^i_{\underset{\vee}{kl}}$ transforms as a tensor. From the point of view of the transformation group it seems therefore at first unnatural to combine these two constituents additively into one single quantity.

3. On the other hand, the lower indices of $\Gamma$ play quite different roles in the defining equation (2) so that there is no compelling reason to restrict the $\Gamma$ by the condition of symmetry with regard to the lower indices. If one does so nevertheless, one is led to the theory of the pure gravitational field. If, however, one does not subject the $\Gamma$ to a restrictive symmetry condition, one arrives at that generalization of the law of gravitation that appears to me as the natural one.

### *The curvature tensor*

Although the $\Gamma$-field does not itself have tensor character, it implies the existence of a tensor. The latter is most easily obtained by displacing a vector $A^i$ according to (2) along the circumference of an infinitesimal two-dimensional surface element and computing its change in one circuit. This change has vector character.

Let $\underset{0}{x^t}$ be the coordinates of a fixed point and $x^t$ those of another point on the circumference. Then $\xi^t = x^t - \underset{0}{x^t}$ is small for all points of the circumference and can be used as a basis for the definition of orders of magnitude.

The integral $\oint \delta A^i$ to be computed is then in more explicit notation

$$-\oint \underline{\Gamma^i_{st}}\, \underline{A^s}\, dx^t \quad \text{or} \quad -\oint \underline{\Gamma^i_{st}}\, \underline{A^s}\, d\xi^t.$$

Underlining of the quantities in the integrand indicates that they are to be taken for successive points of the circumference (and not for the initial point, $\xi^t = 0$).

We first compute in the lowest approximation the value of $\underline{A^i}$ for an arbitrary point $\xi^t$ of the circumference. This lowest approximation is obtained by replacing in the integral, extended now over an open path, $\underline{\Gamma^i_{st}}$ and $\underline{A^s}$ by the values $\Gamma^i_{st}$ and $A^s$ for the initial point of integration ($\xi^t = 0$). The integration gives then

$$\underline{A^i} = A^i - \Gamma^i_{st}\, A^s \int d\xi^t = A^i - \Gamma^i_{st}\, A^s \xi^t.$$

What is neglected here, are terms of second or higher order in $\xi$. With the same approximation one obtains immediately

$$\underline{\Gamma^i_{st}} = \Gamma^i_{st} + \Gamma^i_{st,r}\, \xi^r.$$

Inserting these expressions in the integral above one obtains first, with an appropriate choice of the summation indices,

$$-\oint (\Gamma^i_{st} + \Gamma^i_{st,q}\,\xi^q)\,(A^s - \Gamma^s_{pq}\,A^p\xi^q)\,d\xi^t$$

where all quantities, with the exception of $\xi$, have to be taken for the initial point of integration. We then find

$$-\Gamma^i_{st}\,A^s \oint d\xi^t - \Gamma^i_{st,q}\,A^s \oint \xi^q\,d\xi^t + \Gamma^i_{st}\,\Gamma^s_{pq}\,A^p \oint \xi^q\,d\xi^t$$

where the integrals are extended over the closed circumference. (The first term vanishes because its integral vanishes.) The term proportional to $(\xi)^2$ is omitted since it is of higher order. The two other terms may be combined into

$$[-\Gamma^i_{pt,q} + \Gamma^i_{st}\,\Gamma^s_{pq}]A^p \oint \xi^q\,d\xi^t.$$

This is the change $\Delta A^i$ of the vector $A^i$ after displacement along the circumference. We have

$$\oint \xi^q\,d\xi^t = \oint d(\xi^q\xi^t) - \oint \xi^t\,d\xi^q = -\oint \xi^t\,d\xi^q.$$

This integral is thus antisymmetric in $t$ and $q$, and in addition it has tensor character. We denote it by $f^{tq}$. If $f^{tq}$ were an *arbitrary* tensor, then the vector character of $\Delta A^i$ would imply the tensor character of the bracketed expression in the last but one formula. As it is, we can only infer the tensor character of the bracketed expression if antisymmetrized with respect to $t$ and $q$. This is the *curvature tensor*

$$R^i{}_{klm} \equiv \Gamma^i_{kl,m} - \Gamma^i_{km,l} - \Gamma^i_{sl}\,\Gamma^s_{km} + \Gamma^i_{sm}\,\Gamma^s_{kl}. \qquad (4)$$

The position of all indices is hereby fixed. Contracting

with respect to $i$ and $m$ one obtains the *contracted curvature tensor*

$$R_{ik} \equiv \Gamma^s_{ik,s} - \Gamma^s_{is,k} - \Gamma^s_{it}\,\Gamma^t_{sk} + \Gamma^s_{ik}\,\Gamma^t_{st}. \tag{4a}$$

*The $\lambda$-transformation*

The curvature has a property which will be important in the sequel. For a displacement field $\Gamma$ we may define a new $\Gamma^*$ according to the formula

$$\Gamma^{l}_{ik}{}^* = \Gamma^l_{ik} + \delta^l_i \lambda_{,k} \tag{5}$$

where $\lambda$ is an arbitrary function of the coordinates, and $\delta^l_i$ is the Kronecker tensor ("$\lambda$-transformation"). If one forms $R^i{}_{klm}\,(\Gamma^*)$ by replacing $\Gamma^*$ by the right-hand side of (5), $\lambda$ cancels so that

and
$$\left.\begin{aligned} R^i{}_{klm}\,(\Gamma^*) &= R^i{}_{klm}\,(\Gamma) \\ R_{ik}\;\;(\Gamma^*) &= R_{ik}\;\;(\Gamma) \end{aligned}\right\} \tag{6}$$

The curvature is invariant under $\lambda$-transformations ("$\lambda$-invariance"). Consequently, a theory which contains $\Gamma$ only within the curvature tensor cannot determine the $\Gamma$-field completely but only up to a function $\lambda$, which remains arbitrary. In such a theory, $\Gamma$ and $\Gamma^*$ are to be regarded as representations of the same field, in the same way as if $\Gamma^*$ were obtained from $\Gamma$ merely by a coordinate transformation.

It is noteworthy that the $\lambda$-transformation, contrary to a coordinate transformation, produces a non-symmetric $\Gamma^*$ from a $\Gamma$ that is symmetric in $i$ and $k$. The symmetry condition for $\Gamma$ loses in such a theory its objective significance.

The main significance of $\lambda$-invariance lies in the fact that it has an influence on the "strength" of the system of the field equations, as we shall see later.

# APPENDIX II

*The requirement of "transposition invariance"*

The introduction of non-symmetric fields meets with the following difficulty. If $\Gamma^{l}_{ik}$ is a displacement field, so is $\tilde{\Gamma}^{l}_{ik}$ $(= \Gamma^{l}_{ki})$. If $g_{ik}$ is a tensor, so is $\tilde{g}_{ik}$ $(= g_{ki})$. This leads to a large number of covariant formations among which it is not possible to make a selection on the principle of relativity alone. We shall demonstrate this difficulty by an example and we shall show how it can be overcome in a natural manner.

In the theory of the symmetric field the tensor

$$(W_{ikl} \equiv)\ g_{ik,l} - g_{sk}\,\Gamma^{s}_{il} - g_{is}\,\Gamma^{s}_{lk}$$

plays an important part. If it is put equal to zero, one obtains an equation which permits to express the $\Gamma$ by the $g$, i.e. to eliminate the $\Gamma$. Starting from the facts that (1) $A^{i}{}_{t} \equiv A^{i}{}_{,t} + A^{s}\Gamma^{i}_{st}$ is a tensor, as proved earlier, and that (2) an arbitrary contravariant tensor can be expressed in the form $\sum\limits_{t} \underset{(t)}{A^{i}}\,\underset{(t)}{B^{k}}$, it can be proved without difficulty that the above expression has tensor character also if the fields $g$ and $\Gamma$ are no longer symmetric.

But in the latter case, the tensor character is not lost if, e.g., in the last term $\Gamma^{s}_{lk}$ is transposed, i.e. replaced by $\Gamma^{s}_{lk}$ (this follows from the fact that $g_{is}\,(\Gamma^{s}_{kl} - \Gamma^{s}_{lk})$ is a tensor). There are other formations, though not quite so simple, that conserve the tensor character and can be regarded as extensions of the above expression to the case of the non-symmetric field. Consequently, if one wants to extend to non-symmetric fields the relation between the $g$ and the $\Gamma$ obtained by setting the above expression equal to zero, this seems to involve an arbitrary choice.

But the above formation has a property that distinguishes it from the other possible formations. If one replaces in it simultaneously $g_{ik}$ by $\tilde{g}_{ik}$ and $\Gamma^l_{ik}$ by $\tilde{\Gamma}^l_{ik}$ and then interchanges the indices $i$ and $k$ it is transformed into itself: it is "transposition symmetric" with respect to the indices $i$ and $k$. The equation obtained by putting this expression equal to zero is "transposition invariant." If $g$ and $\Gamma$ are symmetric, this condition is, of course, also satisfied; it is a generalization of the condition that the field quantities be symmetric.

We postulate for the field equations of the non-symmetric field that they be *transposition invariant.* I think that this postulate, physically speaking, corresponds to the requirement that positive and negative electricity enter symmetrically into the laws of physics.

A glance at (4a) shows that the tensor $R_{ik}$ is not completely transposition symmetric, since it transforms by transposition into

$$(R_{ik}{}^* =)\ \Gamma^s_{ik,s} - \Gamma^s_{sk,i} - \Gamma^s_{it}\,\Gamma^t_{sk} + \Gamma^s_{ik}\,\Gamma^t_{ts}\,. \qquad (4b)$$

This circumstance is the basis of the difficulties that one encounters in the endeavour to establish transposition invariant field equations.

### *The pseudo tensor $U^l_{ik}$*

It turns out that a transposition symmetric tensor can be formed from $R_{ik}$ by the introduction of a somewhat different pseudo tensor $U^l_{ik}$ instead of $\Gamma^l_{ik}$. In (4a) the two terms that are linear in $\Gamma$ can be formally combined to a single one. One replaces $\Gamma^s_{ik,s} - \Gamma^s_{is,k}$ by $(\Gamma^s_{ik} - \Gamma^t_{it}\,\delta^s_k)_{,s}$ and defines a new pseudo tensor $U^l_{ik}$ by the equation

$$U^l_{ik} \equiv \Gamma^l_{ik} - \Gamma^t_{it}\,\delta^l_k\,. \qquad (7)$$

Since

$$U^t_{it} = -3\,\Gamma^t_{it}\,,$$

as follows from (7) by contraction with respect to $k$ and $l$, we obtain the following expression for $\Gamma$ in terms of $U$:

$$\Gamma^{l}_{ik} = U^{l}_{ik} - \tfrac{1}{3}\, U^{t}_{it}\, \delta^{l}_{k}\,. \tag{7a}$$

Inserting these in (4a) one finds

$$S_{ik} \equiv U^{s}_{ik,s} - U^{s}_{it}\, U^{t}_{sk} + \tfrac{1}{3}\, U^{s}_{is}\, U^{t}_{tk} \tag{8}$$

for the contracted curvature tensor in terms of $U$. This expression, however, is transposition symmetric. It is this fact that makes the pseudo tensor $U$ so valuable for the theory of non-symmetric fields.

*$\lambda$-transformation for $U$.* If in (5) the $\Gamma$ are replaced by the $U$, one obtains by a simple calculation

$$U^{l}_{ik}{}^{*} = U^{l}_{ik} + (\delta^{l}_{i}\, \lambda_{,k} - \delta^{l}_{k}\, \lambda_{,i})\,. \tag{9}$$

This equation defines the $\lambda$-transformation for the $U$. (8) is invariant with respect to this transformation ($S_{ik}(U^*) = S_{ik}(U)$).

*The transformation law for $U$.* If in (3) and (3a) the $\Gamma$ are replaced by the $U$ with the help of (7a), one obtains

$$U^{l}_{ik}{}^{*} = \frac{\partial x^{l^*}}{\partial x^{l}} \frac{\partial x^{i}}{\partial x^{i^*}} \frac{\partial x^{k}}{\partial x^{k^*}}\, U^{l}_{ik} + \frac{\partial x^{l^*}}{\partial x^{s}} \frac{\partial^2 x^{s}}{\partial x^{i^*} \partial x^{k^*}} - \delta^{l^*}_{k^*} \frac{\partial x^{t^*}}{\partial x^{s}} \frac{\partial^2 x^{s}}{\partial x^{i^*} \partial x^{t^*}}\,. \tag{10}$$

Note that again the indices referring to both systems assume all the values from 1 to 4 *independently of each other*, even though the same letter is being used. Regarding this formula it is noteworthy that on account of the last term it is not transposition symmetric with respect to the indices $i$ and $k$. This peculiar circumstance can be clarified by demonstrating that this transformation may be regarded as a composition of a transposition symmetric coordinate transformation and a $\lambda$-transformation. In order to see that we write first the last term in the form

$$-\tfrac{1}{2}\left[\delta_{k^*}^{l^*}\frac{\partial x^{t^*}}{\partial x^s}\frac{\partial^2 x^s}{\partial x^{i^*}\partial x^{t^*}}+\delta_{i^*}^{l^*}\frac{\partial x^{t^*}}{\partial x^s}\frac{\partial^2 x^s}{\partial x^{k^*}\partial x^{t^*}}\right]$$
$$+\tfrac{1}{2}\left[\delta_{i^*}^{l^*}\frac{\partial x^{t^*}}{\partial x^s}\frac{\partial^2 x^s}{\partial x^{k^*}\partial x^{t^*}}-\delta_{k^*}^{l^*}\frac{\partial x^{t^*}}{\partial x^s}\frac{\partial^2 x^s}{\partial x^{i^*}\partial x^{t^*}}\right].$$

The first of these two terms is transposition symmetric. Let us combine it with the first two terms of the right-hand side of (10) to an expression $K_{ik}^{l}{}^*$. Let us now consider what we get if the transformation

$$U_{ik}^{l}{}^* = K_{ik}^{l}{}^*$$

is followed by the $\lambda$-transformation

$$U_{ik}^{l}{}^{**} = U_{ik}^{l}{}^* + \delta_{i^*}^{l^*}\lambda_{,k^*} - \delta_{k^*}^{l^*}\lambda_{,i^*}.$$

The composition yields

$$U_{ik}^{l}{}^{**} = K_{ik}^{l}{}^* + (\delta_{i^*}^{l^*}\lambda_{,k^*} - \delta_{k^*}^{l^*}\lambda_{,i^*}).$$

This implies that (10) may be regarded as such a composition provided the second term of (10a) can be brought into the form $\delta_{i^*}^{l^*}\lambda_{,k^*} - \delta_{k^*}^{l^*}\lambda_{,i^*}$. For this it is sufficient to show that a $\lambda$ exists such that

$$\tfrac{1}{2}\frac{\partial x^{t^*}}{\partial x^s}\frac{\partial^2 x^s}{\partial x^{k^*}\partial x^{t^*}} = \lambda_{,k^*} \tag{11}$$

$$\left(\text{and } \tfrac{1}{2}\frac{\partial x^{t^*}}{\partial x^s}\frac{\partial^2 x^s}{\partial x^{i^*}\partial x^{t^*}} = \lambda_{,i^*}\right).$$

In order to transform the left-hand side of the so far hypothetical equation we must first express $\frac{\partial x^{t^*}}{\partial x^s}$ by the coefficients of the inverse transformation, $\frac{\partial x^a}{\partial x^{b^*}}$. On the one hand,

$$\frac{\partial x^p}{\partial x^{t^*}}\frac{\partial x^{t^*}}{\partial x^s} = \delta_s^p. \tag{a}$$

On the other,

$$\frac{\partial x^p}{\partial x^{t^*}} V^s_{t^*} = \frac{\partial x^p}{\partial x^{t^*}} \frac{\partial D}{\partial \left(\dfrac{\partial x^s}{\partial x^t}\right)} = D\delta^p_s.$$

Here, $V^s_{t^*}$ denotes the co-factor of $\dfrac{\partial x^s}{\partial x^{t^*}}$, and may in turn be expressed as the derivative of the determinant $D = \left|\dfrac{\partial x^a}{\partial x^{b^*}}\right|$ with respect to $\dfrac{\partial x^s}{\partial x^{t^*}}$. Therefore, we have also

$$\frac{\partial x^p}{\partial x^{t^*}} \cdot \frac{\partial \log D}{\partial \left(\dfrac{\partial x^s}{\partial x^{t^*}}\right)} = \delta^p_s. \tag{b}$$

It follows from (a) and (b) that

$$\frac{\partial x^{t^*}}{\partial x^s} = \frac{\partial \log D}{\partial \left(\dfrac{\partial x^s}{\partial x^{t^*}}\right)}.$$

Because of this relation the left-hand side of (11) can be written as

$$\tfrac{1}{2} \frac{\partial \log D}{\partial \left(\dfrac{\partial x^s}{\partial x^{t^*}}\right)} \left(\frac{\partial x^s}{\partial x^{t^*}}\right)_{,k^*} = \tfrac{1}{2} \frac{\partial \log D}{\partial x^{k^*}}.$$

This implies that (11) is indeed satisfied by

$$\lambda = \tfrac{1}{2} \log D.$$

This proves that the transformation (10) can be regarded as a composition of the transposition symmetric transformation

$$U^l_{ik}{}^* = \frac{\partial x^{l^*}}{\partial x^l} \frac{\partial x^i}{\partial x^{i^*}} \frac{\partial x^k}{\partial x^{k^*}} U^l_{ik} + \frac{\partial x^{l^*}}{\partial x^s} \frac{\partial^2 x^s}{\partial x^{i^*} \partial x^{k^*}}$$
$$- \tfrac{1}{2} \left[\delta^{l^*}_{k^*} \frac{\partial x^{t^*}}{\partial x^s} \frac{\partial^2 x^s}{\partial x^{i^*} \partial x^{t^*}} + \delta^{l^*}_{i^*} \frac{\partial x^{t^*}}{\partial x^s} \frac{\partial^2 x^s}{\partial x^{k^*} \partial x^{t^*}}\right] \tag{10b}$$

and a $\lambda$-transformation. (10b) may thus be taken in place of (10) as transformation formula for the $U$. Any transformation of the $U$-field that only changes the *form* of the representation can be expressed as a composition of a coordinate transformation according to (10b) and a $\lambda$-transformation.

### *Variational principle and field equations*

The derivation of the field equations from a variational principle has the advantage that the compatibility of the resulting system of equations is assured and that the identities connected with the general covariance, the "Bianchi identities," as well as the conservation laws result in a systematic manner.

The integral to be varied requires as integrand $\mathfrak{H}$ a scalar density. We shall construct such a density from $R_{ik}$ or $S_{ik}$. The simplest procedure is to introduce a covariant tensor density $\mathfrak{g}^{ik}$ of weight 1 in addition to $\Gamma$ or $U$ respectively, setting

$$\mathfrak{H} = \mathfrak{g}^{ik} R_{ik} \; (= \mathfrak{g}^{ik} S_{ik}). \tag{12}$$

The transformation law for the $\mathfrak{g}^{ik}$ must be

$$\mathfrak{g}^{ik^*} = \frac{\partial x^{i^*}}{\partial x^i} \frac{\partial x^{k^*}}{\partial x^k} \mathfrak{g}^{ik} \left| \frac{\partial x^t}{\partial x^{t^*}} \right| \tag{13}$$

where again the indices referring to different coordinate systems, in spite of the use of the same letters, are to be treated as independent of each other. We obtain indeed

$$\int \mathfrak{H}^* \, d\tau^* = \int \frac{\partial x^{i^*}}{\partial x^i} \frac{\partial x^{k^*}}{\partial x^k} \mathfrak{g}^{ik} \left| \frac{\partial x^t}{\partial x^{t^*}} \right| \cdot \frac{\partial x^s}{\partial x^{i^*}} \frac{\partial x^t}{\partial x^{k^*}} S_{st} \left| \frac{\partial x^{r^*}}{\partial x^r} \right| d\tau$$

$$= \int \mathfrak{H} \, d\tau,$$

i.e. the integral is transformation invariant. Furthermore, the integral is invariant with respect to a $\lambda$-transformation (5) or (9) because $R_{ik}$ as expressed by the $\Gamma$ or $U$ respectively, and hence also $\mathfrak{H}$, is invariant with respect to a $\lambda$-transformation. From this it follows that also the field equations to be derived by variation of $\int \mathfrak{H} d\tau$ are covariant with respect to coordinate and to $\lambda$-transformations.

But we also postulate that the field equations are to be transposition invariant with respect to the two fields $\mathfrak{g}$, $\Gamma$ or the fields $\mathfrak{g}$, $U$. This is assured if $\mathfrak{H}$ is transposition invariant. We have seen that $R_{ik}$ is transposition symmetric if expressed in the $U$, but not if expressed in the $\Gamma$. Hence $\mathfrak{H}$ is only transposition invariant if we introduce in addition to the $\mathfrak{g}^{ik}$ the $U$ (but not the $\Gamma$) as field variables. In that case, we are sure from the beginning that the field equations derived from $\int \mathfrak{H}\, d\tau$ by variation of the field variables are transposition invariant.

By variation of $\mathfrak{H}$ (equations (12) and (8)) with respect to the $\mathfrak{g}$ and $U$ we find

$$\left.\begin{aligned} \delta\mathfrak{H} &= S_{ik}\,\delta\mathfrak{g}^{ik} - \mathfrak{N}^{ik}{}_l\,\delta U^l_{ik} + (\mathfrak{g}^{ik}\,\delta U^s_{ik})_{,s} \\ \text{where}\quad S_{ik} &= U^s_{ik,s} - U^s_{it}\,U^t_{sk} + \tfrac{1}{3}\,U^s_{is}\,U^t_{tk}, \\ \mathfrak{N}^{ik}{}_l &= \mathfrak{g}^{ik}{}_{,l} + \mathfrak{g}^{sk}\,(U^i_{sl} - \tfrac{1}{3}\,U^t_{st}\,\delta^i_l) \\ &\quad + \mathfrak{g}^{is}\,(U^k_{ls} - \tfrac{1}{3}\,U^t_{ts}\,\delta^k_l). \end{aligned}\right\} \qquad (14)$$

*The field equations*

Our variational principle is

$$\delta\left(\int \mathfrak{H}\, d\tau\right) = 0 \qquad (15)$$

# APPENDIX II

The $\mathfrak{g}^{ik}$ and $U^l_{ik}$ are to be varied independently, their variations vanishing at the boundary of the domain of integration. This variation gives first of all

$$\int \delta\mathfrak{H}\, d\tau = 0.$$

If the expression given in (14) is inserted here, the last term of the expression for $\delta\mathfrak{H}$ does not give any contribution since $\delta U^l_{ik}$ vanishes at the boundary. Hence we obtain the field equations

$$S_{ik} = 0 \tag{16a}$$

$$\mathfrak{N}^{ik}{}_l = 0. \tag{16b}$$

They are—as is already evident from the choice of the variational principle—invariant with respect to coordinate and to $\lambda$-transformations and also transposition invariant.

*Identities*

These field equations are not independent of each other. Between them exist $4 + 1$ identities. That is, there exist $4 + 1$ equations between their left-hand sides that hold regardless of whether or not the $\mathfrak{g}$-$U$ field satisfies the field equations.

These identities can be derived by a well-known method from the fact that $\int \mathfrak{H}\, d\tau$ is invariant with respect to coordinate and to $\lambda$-transformations.

For it follows from the invariance of $\int \mathfrak{H}\, d\tau$ that its variation vanishes *identically* if one inserts in $\delta\mathfrak{H}$ the variations $\delta\mathfrak{g}$ and $\delta U$ which arise from an infinitesimal coordinate transformation or an infinitesimal $\lambda$-transformation respectively.

# APPENDIX II

An infinitesimal coordinate transformation is described by

$$x^{i^*} = x^i + \xi^i \tag{17}$$

where $\xi^i$ is an arbitrary infinitesimal vector. We must now express the $\delta \mathfrak{g}^{ik}$ and $\delta U^l_{ik}$ by the $\xi^i$ using the equations (13) and (10b). Because of (17) one must replace

$$\frac{\partial x^{a^*}}{\partial x^b} \text{ by } \delta^a_b + \xi^a{}_{,b}, \quad \frac{\partial x^a}{\partial x^{b^*}} \text{ by } \delta^a_b - \xi^a{}_{,b}$$

and omit all terms that are of higher than first order in $\xi$. Thus one obtains

$$\delta \mathfrak{g}^{ik} (= \mathfrak{g}^{ik^*} - \mathfrak{g}^{ik}) = \mathfrak{g}^{sk} \xi^i{}_{,s} + \mathfrak{g}^{is} \xi^k{}_{,s} - \mathfrak{g}^{ik} \xi^s{}_{,s} + [-\mathfrak{g}^{ik}{}_{,s} \xi^s] \tag{13a}$$

$$\delta U^l_{ik} (= U^l_{ik}{}^* - U^l_{ik}) = U^s_{ik} \xi^l{}_{,s} - U^l_{sk} \xi^s{}_{,i} - U^l_{is} \xi^s{}_{,k} + \xi^l{}_{,ik} + [- U^l{}_{ik,s} \xi^s]. \tag{10c}$$

Note here the following. The transformation formulas furnish the new values of the field variables *for the same point of the continuum.* The calculation indicated above first gives expressions for $\delta \mathfrak{g}^{ik}$ and $\delta U^l_{ik}$ without the terms in brackets. In the calculus of variation, on the other hand, $\delta \mathfrak{g}^{ik}$ and $\delta U^l_{ik}$ denote the variations *for fixed values of the coordinates.* In order to obtain these the terms in brackets have to be added.

If one inserts in (14) these "transformation variations" $\delta \mathfrak{g}$ and $\delta U$, the variation of the integral $\int \mathfrak{H}\, d\tau$ vanishes identically. If furthermore the $\xi^i$ are so chosen that they vanish together with their first derivatives at the boundary of the domain of integration, the last term in (14) gives no contribution. The integral

$$\int (S_{ik}\, \delta \mathfrak{g}^{ik} - \mathfrak{N}^{ik}{}_l\, \delta U^l_{ik})\, d\tau$$

vanishes therefore identically if the $\delta \mathfrak{g}^{ik}$ and $\delta U^l_{ik}$ are replaced by the expressions (13a) and (10c). Since this integral depends linearly and homogeneously on the $\xi^i$ and their derivatives, it can be brought into the form

$$\int \mathfrak{W}_i \, \xi^i \, d\tau$$

by repeated integration by parts, where $\mathfrak{W}_i$ is a known expression (of first order in $S_{ik}$ and of second order in $\mathfrak{N}^{ik}{}_l$). From this follow the identities

$$\mathfrak{W}_i \equiv 0. \tag{18}$$

These are four identities for the left-hand sides $S_{ik}$ and $\mathfrak{N}^{ik}{}_l$ of the field equations, which correspond to the Bianchi identities. According to the terminology introduced before these identities are of third order.

There exists a fifth identity corresponding to the invariance of the integral $\int \mathfrak{H} \, d\tau$ with respect to infinitesimal $\lambda$-transformations. Here we have to insert in (14)

$$\delta \mathfrak{g}^{ik} = 0 \qquad \delta U^l_{ik} = \delta^l_i \, \lambda_{,k} - \delta^l_k \, \lambda_{,i}$$

where $\lambda$ is infinitesimal and vanishes at the boundary of the domain of integration. One obtains first

$$\int \mathfrak{N}^{ik}{}_l \, (\delta^l_i \, \lambda_{,k} - \delta^l_k \, \lambda_{,i}) \, d\tau = 0$$

or, after integration by parts,

$$2 \int \mathfrak{N}^{is}_{\smile \, s,i} \, \lambda \, d\tau = 0$$

(where, generally, $\mathfrak{N}^{ik}_{\smile \, l} = \frac{1}{2} \, (\mathfrak{N}^{ik}{}_l - \mathfrak{N}^{ki}{}_l)$).

This furnishes the desired identity

$$\mathfrak{N}^{\underset{\vee}{is}}{}_{s,i} \equiv 0. \tag{19}$$

In our terminology this is an identity of second order. For $\mathfrak{N}^{\underset{\vee}{is}}{}_{s}$ we obtain from (14) by straightforward computation

$$\mathfrak{N}^{\underset{\vee}{is}}{}_{s} \equiv \mathfrak{g}^{\underset{\vee}{is}}{}_{,s}\,. \tag{19a}$$

If the field equation (16b) is satisfied, we have thus

$$\mathfrak{g}^{\underset{\vee}{is}}{}_{,s} = 0. \tag{16c}$$

*Remark on the physical interpretation.* A comparison with Maxwell's theory of the electromagnetic field suggests the interpretation that (16c) expresses the vanishing of the magnetic current density. If this is accepted, it is evident which expression should denote the electric current density. One can assign a tensor $g^{ik}$ to the tensor density $\mathfrak{g}^{ik}$ by setting

$$\mathfrak{g}^{ik} = g^{ik}\sqrt{-|g_{st}|} \tag{20}$$

where the covariant tensor $g_{ik}$ is correlated to the contravariant one by the equations

$$g_{is}\,g^{ks} = \delta_i^k. \tag{21}$$

From these two equations we obtain

$$g^{ik} = \mathfrak{g}^{ik}\,(-|\mathfrak{g}^{st}|)^{-\frac{1}{2}}$$

and then $g_{ik}$ from equations (21). We may then assume that

$$(a_{ikl}) = g_{\underset{\vee}{ik},l} + g_{\underset{\vee}{kl},i} + g_{\underset{\vee}{li},k} \tag{22}$$

or

$$\mathfrak{a}^m = \tfrac{1}{6}\,\eta^{iklm}\,a_{ikl} \tag{22a}$$

expresses the current density, where $\eta^{iklm}$ is Levi-Cività's tensor density (with components $\pm 1$) antisymmetric in all indices. The divergence of this quantity vanishes identically.

## APPENDIX II

*The strength of the system of equations* (16*a*), (16*b*)

In applying here the method of enumeration described above one must take into account the fact that all the $U^*$ obtained from a given $U$ by $\lambda$-transformations of the form (9) actually represent the same $U$-field. This has the consequence that the $n$th order coefficients of the $U^l_{ik}$-expansion incorporate $\binom{4}{n}$ $n$th order derivatives of $\lambda$ whose choice is of no consequence for the distinction of actually differing $U$-fields. Thus the number of expansion coefficients relevant for the enumeration of the $U$-fields is decreased by $\binom{4}{n}$. By the enumeration method we obtain for the number of free $n$th order coefficients

$$z = \left[16\binom{4}{n} + 64\binom{4}{n-1} - 4\binom{4}{n+1} - \binom{4}{n}\right] - \left[16\binom{4}{n-2} + 64\binom{4}{n-1}\right] + \left[4\binom{4}{n-3} + \binom{4}{n-2}\right]. \tag{23}$$

The first bracket represents the total number of relevant $n$th order coefficients which characterize the g-$U$-field, the second the reduction of this number due to the existence of the field equations, and the third bracket gives the correction to this reduction on account of the identities (18) and (19). Computing the asymptotic value for large $n$ we find

$$z \sim \binom{4}{n}\frac{z_1}{n} \tag{23a}$$

where

$$z_1 = 42.$$

The field equations of the non-symmetric field are thus considerably weaker than those of the pure gravitational field ($z_1 = 12$).

*The influence of λ-invariance on the strength of the system of equations.* One may be tempted to bring about transposition invariance of the theory by starting from the transposition invariant expression

$$\mathfrak{H} = \tfrac{1}{2}(\mathfrak{g}^{ik} R_{ik} + \tilde{\mathfrak{g}}^{ik} \tilde{R}_{ik})$$

(instead of introducing the $U$ as field variables). Of course, the resulting theory will be different from the one expounded above. It can be shown that for this $\mathfrak{H}$ no λ-invariance exists. Here, too, we obtain field equations of the type (16a), (16b), which are transposition invariant (with respect to $\mathfrak{g}$ and $\Gamma$). Between them, there exist, however, only the four "Bianchi identities." If one applies the method of enumeration to this system, then, in the formula corresponding to (23), the fourth term in the first bracket and the second term in the third bracket are missing. One obtains

$$z_1 = 48.$$

The system of equations is thus weaker than the one chosen by us and is therefore to be rejected.

*Comparison with the previous system of field equations.* This is given by

$$\Gamma^s_{\underset{\vee}{is}} = 0 \qquad\qquad R_{\underline{ik}} = 0$$

$$g_{ik,l} - g_{sk}\,\Gamma^s_{il} - g_{is}\,\Gamma^s_{lk} = 0 \qquad R_{\underset{\vee}{ik},l} + R_{\underset{\vee}{kl},i} + R_{\underset{\vee}{li},k} = 0$$

where $R_{ik}$ is defined by (4a) as a function of the $\Gamma$ (and where $R_{\underline{ik}} = \tfrac{1}{2}(R_{ik} + R_{ki})$, $R_{\underset{\vee}{ik}} = \tfrac{1}{2}(R_{ik} - R_{ki})$).

This system is entirely equivalent to the new system

(16a), (16b) since it has been derived from the same integral by variation. It is transposition invariant with respect to the $g_{ik}$ and $\Gamma^l_{ik}$. The difference, however, lies in the following. The integral to be varied is itself not transposition invariant, nor is the system of equations that is at first obtained by its variation; it is, however, invariant with respect to the $\lambda$-transformations (5). In order to obtain transposition invariance here, one has to use an artifice. One formally introduces four new field variables, $\lambda_i$, which after variation are so chosen that the equations $\Gamma^s_{\underset{\vee}{is}} = 0$ are satisfied.* Thus the equations obtained by variation with respect to the $\Gamma$ are brought into the indicated transposition invariant form. But the $R_{ik}$-equations still contain the auxiliary variables $\lambda_i$. One can, however, eliminate them, which leads to a decomposition of these equations in the manner stated above. The equations obtained are then also transposition invariant (with respect to the $\mathfrak{g}$ and $\Gamma$).

Postulating the equations $\Gamma^s_{\underset{\vee}{is}} = 0$ involves a normalization of the $\Gamma$-field, which removes the $\lambda$-invariance of the system of equations. As a result, not all equivalent representations of a $\Gamma$-field appear as solutions of this system. What takes place here, is comparable to the procedure of adjoining to the field equations of pure gravitation arbitrary additional equations which restrict the choice of coordinates. In our case, moreover, the system of equations becomes unnecessarily complicated. These difficulties are avoided in the new representation by starting from a variational principle that is transposition invariant with respect to the $\mathfrak{g}$ and $U$, and by using throughout the $\mathfrak{g}$ and $U$ as field ariables.

* By setting $\Gamma^l_{ik}{}^* = \Gamma^l_{ik} + \delta^l_i \lambda_k$.

# APPENDIX II

*The divergence law and the conservation law of momentum and energy*

If the field equations are satisfied and if, moreover, the variation is a transformation variation, then, in (14), not only $S_{ik}$ and $\mathfrak{N}^{ik}{}_l$ vanish, but also $\delta\mathfrak{H}$, so that the field equations imply the equations

$$(\mathfrak{g}^{ik}\,\delta U^{s}_{ik})_{,s} = 0$$

where $\delta U^{s}_{ik}$ is given by (10c). This divergence law holds for any choice of the vector $\xi^i$. The simplest special choice, i.e. $\xi^i$ independent of the $x$, leads to the four equations

$$\mathfrak{T}^{s}_{t,s} \equiv (\mathfrak{g}^{ik}\, U^{s}_{ik,t})_{,s} = 0.$$

These can be interpreted and applied as the equations of conservation of momentum and energy. It should be noted that such conservation equations are never uniquely determined by the system of field equations. It is interesting that according to the equations

$$\mathfrak{T}^{s}_{t} \equiv \mathfrak{g}^{ik}\, U^{s}_{ik,t}$$

the density of the energy current $(\mathfrak{T}^1_4, \mathfrak{T}^2_4, \mathfrak{T}^3_4)$ as well as the energy density $\mathfrak{T}^4_4$ vanish for a field that is independent of $x^4$. From this one can conclude that according to this theory a stationary field free from singularities can never represent a mass different from zero.

The derivation as well as the form of the conservation laws become much more complicated if the former formulation of the field equations is used.

### GENERAL REMARKS

A. In my opinion the theory presented here is the logically simplest relativistic field theory which is at all possible. But this does not mean that nature might not obey a more complex field theory.

More complex field theories have frequently been proposed. They may be classified according to the following characteristic features:

(a) Increase of the number of dimensions of the continuum. In this case one must explain why the continuum is *apparently* restricted to four dimensions.

(b) Introduction of fields of a different kind (e.g. a vector field) in addition to the displacement field and its correlated tensor field $g_{ik}$ (or $\mathfrak{g}^{ik}$).

(c) Introduction of field equations of higher order (of differentiation).

In my view, such more complicated systems and their combinations should be considered only if there exist physical-empirical reasons to do so.

B. A field theory is not yet completely determined by the system of field equations. Should one admit the appearance of singularities? Should one postulate boundary conditions? As to the first question, it is my opinion that singularities must be excluded. It does not seem reasonable to me to introduce into a continuum theory points (or lines etc.) for which the field equations do not hold. Moreover, the introduction of singularities is equivalent to postulating boundary conditions (which are arbitrary from the point of view of the field equations) on "surfaces" which closely surround the singularities. Without such a postulate the theory is much too vague. In my opinion the answer to the second question is that the postulation of boundary conditions is indispensable. I shall demonstrate this by an elementary example. One can compare the postulation of a potential of the form $\phi = \sum \frac{m}{r}$ with the statement that outside the mass points (in three dimensions) the equation $\Delta\phi = 0$ is satisfied. But if one does not add the boundary condition that $\phi$ vanish (or remain finite) at infinity, then there exist solutions that are entire functions of the $x$ (e.g. $x_1^2 - \frac{1}{2}(x_2^2 + x_3^2)$) and become infinite at infinity. Such fields can only be excluded by postulating a boundary condition in case the space is an "open" one.

C. Is it conceivable that a field theory permits one to understand the atomistic and quantum structure of reality? Almost everybody will answer this question with "no." But I believe that at the present time nobody knows anything reliable about it. This is so because we cannot judge in what manner and how strongly the exclusion of singularities reduces the manifold of solutions. We do not possess any method at all to derive systematically solutions that are free of singularities. Approximation methods are of no avail since one never knows whether or not there exists to a particular approximate solution an exact solution *free of singularities*. For this reason we cannot at present compare the content of a nonlinear field theory with experience. Only a significant progress in the mathematical methods can help here. At the present time the opinion prevails that a field theory must first, by "quantization," be transformed into a statistical theory of field probabilities according to more or less established rules. I see in this method only an attempt to describe relationships of an essentially nonlinear character by linear methods.

D. One can give good reasons why reality cannot at all be represented by a continuous field. From the quantum phenomena it appears to follow with certainty that a finite system of finite energy can be completely described by a finite set of numbers (quantum numbers). This does not seem to be in accordance with a continuum theory, and must lead to an attempt to find a purely algebraic theory for the description of reality. But nobody knows how to obtain the basis of such a theory.

# INDEX

# INDEX

PROPERTY OF THE LIBRARY
ASSUMPTION COLLEGE
BRANTFORD, ONTARIO